TELEPATHY our LOST SENSE

Neuroscience sheds Light on ESP

by
Dr Dianne Cartwright

Telepathy our Lost Sense
Copyright © 2024 by Dianne Cartwright

All rights reserved. No part of this publication may be reproduced, distributed, or transmitted by any person or entity, including internet search engines or retailers, in any form or by any means, electronic or mechanical including photocopying (except under the statutory provision exceptions of the Australian copyright act 1968) recording, scanning or by any information storage and retrieval system without the prior written permission of the publisher/author Dianne Cartwright.

ISBN: 978-1-7637669-0-7
Published by Dianne Cartwright
diannecartwright.com.au

Typsetting by Rack and Rune Publishing
rackandrune.com
email: info@rackandrune.com
Cover Design
Mind Mob Studio
website: mindmob.com.au

For my patients who so willingly shared their experiences with me

"Nothing in life is to be feared, it is only to be understood. Now is the time to understand more, so that we may fear less."

Marie Curie – Nobel Laureate, scientist, discoverer of radium, wife and mother

Acknowledgements

First and foremost, I wish to thank the patients who so willingly and generously took part in the survey, giving me their trust and time, and the assistance of my receptionist Deborah C who kept them happy when I was running late listening to their experiences.

Secondly, I wish to thank my husband Bruce who patiently provided encouragement, numerous cups of tea and coffee and read the manuscript many times through its various iterations. He provided valuable assistance with corrections turning my upside down, inside out and fractured English into readable language.

To all my good friends who read the manuscript and offered constructive advice, criticism and encouragement – Anne A, Sharon O, Oonagh P, and David S - a big thank you and especially my late friend Mike Pemberton who assisted with the physics.

Finally, to those who helped transition the manuscript into the book –

Kit Carstairs from the Manuscript Appraisal Agency and her team, the structural and copy editors who are anonymous, but greatly appreciated. Dr Susan Jacups - statistician Cairns, who provided advice and oversight of the statistical analysis of the results of the survey. Liesl Krummey and her team from Mindmob Studio who interpreted and executed my vision for the book cover perfectly. Graham Davidson of RackandRune for styling, presentation, formatting and invaluable advice. Lastly to Angela Hoskins of Site Content Web Design for my amazing website.

Contents

Part I: The Facts
    Telepathy – My Experiences   15
    What are Esp and Telepathy?   18
    Telepathy or Coincidence?   21
    The Survey   25
    Telepathic Experiences   37
    The Nature of Telepathy   45

Part II: Science of the Mind
    Electromagnetic Radiation and our Senses   53
    Neurones and Brainwaves   58
    The Human Brain   65
    Coding Thoughts   68
    Electrical Signals bypass the Normal Senses   74
    A New Sense – the Magnetic Field   80
    Is the Human Brain an Antenna?   87
    Interpreting Telepathic Information   95
    The Hypothesis in a Nutshell   105

Part III: Mysterious Extrasensory Perception
    Visitations from the Dying   110
    Brainwaves and ESP Phenomena   119

Part IV: Distortions of Time
    Precognition: Seeing the Future   126
    Dreaming the Future   139
    Ghosts   148
    Touchy Feely-Ghosts   155
    What is Wrong with Time?   158
    Where does God fit in?   167
    The Final Word   169

Appendix
    Section 1: Consent, Survey Questions & Results   171
    Section 2: The Experiences   181
    Section 3: Calculations of Word and Timing Coincidence   183
    Section 4: Diagrams   185

References   **188**

Preface

Telepathy and extrasensory perception (ESP) intrigue millions of people all over our world. Surveys show that belief in these phenomena is widespread and stories concerning them permeate our culture in books, film, and television. However, people rarely talk about their personal experiences for fear of ridicule or worse. In previous centuries, those with abilities in these areas faced accusations of witchcraft, punishment and death. Until recently, Wikipedia had a derogatory definition of telepathy – calling it the purported messaging of thoughts, equating it with "hearing voices" and suggesting that it was symptomatic of a severe mental illness or brain disease, such as epilepsy. Unfortunately, disclosing an interest in these topics in the world of scientists is taboo and funding for research in this area is virtually non-existent.

My curiosity was sparked by my own experiences, which commenced in my teenage years, and those of friends who confided in me. While my personal experiences have been few and far between, they are incredible and unforgettable. As I became more comfortable discussing this topic with friends and hearing their equally amazing stories, I asked myself two very simple questions: *How many people have these experiences? Were we odd or was this a normal part of the human experience?*

To find the answers, I surveyed the patients attending my general practice in Cairns, a coastal town in far north Queensland, Australia. Almost 170 patients were eligible to take part in the survey and the results were surprising. Just under fifty percent of them had at least one extrasensory experience and over seventy-five percent had received or sent a telepathic message to one other person. This astonishing number of positive responses indicates that telepathy and extrasensory perception were common in my patient population. In medicine, a common illness occurs in ten percent of a population while a rare illness is present in fewer than one in 1000 people. This suggests that telepathy and extrasensory perception are natural human abilities and should not be thought of as rare or a manifestation of illness.

The patients shared over 200 experiences with me, ranging from straight forward telepathy to "visits" from dying loved ones, seeing the future while awake or in dreams, and seeing or feeling ghosts, amongst other things.

There must be an explanation for this ability – a biological process involved in the reception of information from one individual to another without the use of the "normal" senses of sight, hearing or touch. Furthermore, this process must be rooted in our physiology and conform with the laws of physics, like our "normal" senses.

One of my scientific heroes is William Harvey, who discovered the circulation of the blood and published his work in 1628. By observing the flow of blood in the veins towards the heart and the volume of blood in the left ventricle of the heart, he was able to calculate the amount of blood pumped out by the heart in half a minute, thereby deducing that the blood must circulate around the body and back to the heart. It took another two centuries, the invention of the microscope and the discovery of capillaries connecting the veins and arteries, before scientists proved Harvey's hypothesis.

In collecting and analysing an immense amount of data from the survey, I aspired to emulate Harvey and shed some light on how a person can receive and send telepathic messages. For many years, research into ESP consisted of collecting anecdotes from interested people and conducting guessing games using Zener cards. These experiments were severely criticised for faulty design and improper use of statistics, not to mention possible fraud. Later research used more rigorous methodology, such as the Ganzfield experiments, where participants' sight and hearing were shielded while attempting to guess which card of five possibilities had been chosen.[1,2] While some of the results were statistically significant, they were hardly likely to set the world on fire. It is not awe-inspiring to show that participants choosing between two options correctly guessed fifty-three percent of the time – instead of fifty percent.

Research into ESP has languished in the doldrums since the 1990s. With one or two rare exceptions, no serious experiments have been carried out since then. The US government provided funding through the CIA for experimentation with "remote viewing" by some adepts and psychics in the 1970s and 80s at the Stanford Research Institute in California. Edgar Mitchell, one of the astronauts on the Apollo missions, was a founder of the Noetic Sciences Institute, which conducted research into some aspects of presentiment – that is, "feeling the future". Psychologist Daniel Bem was soundly criticised for his research into precognition – using students who preferentially remembered words from a list that they would later learn in the future. Finally, at the beginning of the twenty-first century, the late Dr Michael Persinger – a psychologist – and his coworkers at Laurentian College in Canada studied the brainwaves of an expert while he was "mind reading."

I took my two signature stories of a telepathic message – one I had received and one I had sent – to a friend who is a quantum physicist. She listened patiently and then said, "I didn't think I would be able to help you, but I think your brain is acting as an antenna for telepathic messages carried by electromagnetic radiation." This had been proposed in the past by a Russian researcher and by Dr Persinger but was unable to be developed for lack of a plausible mechanism.

Now, exciting new knowledge is available that enabled me to research my friend's idea. In 2013, President Barack Obama announced the BRAIN Initiative (Brain Research through Advancing Innovative Neurotechnologies), designed to investigate how the brain processes information into thoughts and consequent behaviour. The BRAIN initiative has paid off in myriad ways and has filled many gaps in our understanding of the brain and its neurones.

We have developed new technology with quantitative electroencephalographs (EEGs) and functional magnetic resonance imaging (fMRI scans) to study the working brain and designate the areas and pathways the brain uses – for example while doing mental arithmetic. Lastly and most importantly, Brain–Computer-Interfacing (BCI) can transmit thoughts from one brain to a computer and can also facilitate transmission between one human mind and another. Also, a whole new world of research has described the interaction of Earth's magnetic field and living organisms, to prove that many animals have a magnetic sense.

It follows that natural telepathy without the interposition of a computing system is theoretically possible.

The first section of the book discusses the nature of telepathy, details my signature telepathic messaging, and tells of some of the representative telepathic experiences of my patients. The results of the survey are analysed – with a particular focus on those showing that patients with mental illnesses are no more or less likely than other people to have ESP experiences or telepathic abilities. Some of the qualities and characteristics of telepathic messages are derived and discussed in this section.

The second part of the book delves into the science of electromagnetic radiation, the neuroanatomy and neurophysiology of the brain and its billions of neurones, and how these function to create electrical and magnetic brainwaves which carry information in the form of electric or magnetic codes from one area of the brain to another. Brain– computer interfacing and artificial telepathy are discussed here.

New discoveries in neuroscience, earth geomagnetic science and some of Dr Persinger's studies enabled the development of a hypothesis which explains the type of electromagnetic radiation most likely to carry telepathic information between people and the role of the brain in receiving, decoding and transmitting these messages to and from other minds.

The third section of the book applies this hypothesis and reviews recent discoveries related to the common ESP experience of visualising or knowing of a person's death at the time of their passing. Evidence for brain activity during other ESP experiences such as psychokinesis and out-of-body experiences is presented. The patients described many examples of seeing or knowing the future – commonly known as clairvoyance – and the history of research into these phenomena is discussed, including that on laboratory animals.

Several patients reported seeing or feeling ghosts. Both clairvoyance and viewing ghosts cast doubt on our widely accepted beliefs about time. Time is supposed to proceed in an orderly manner from the past to the present to the future, and we are supposed to live and act only in the present. But if this is truly the case, seeing the future is impossible. The flow of time is a controversial topic in physics and philosophy and melds into theories concerning the nature and existence of the Universe. If we accept that these experiences of the future are real, then our ideas about time need to change.

My dream in writing this book is to make this wonderful story of our brain's ability to communicate with another person at a distance available to all humankind. It is a dream to relieve fear of these abilities, to open the eyes of those who have no personal experiences of this nature, to enlighten the sceptics and to encourage the scientists to pursue and investigate this old and hidden resource we possess.

To share this vision with me, read on!

Part I: The Facts

1

Telepathy – My Experiences

My middle years were very busy with establishing my general practice, raising children, spending time with friends and family, and taking time out on the weekends. I was working in my garden on a quiet Sunday morning – my mind blank as I pulled out weeds and pruned – when a thought appeared from nowhere: *Anna is coming to see me tomorrow morning.* "Whaaat?" I thought. "Where did that come from?" Anna was one of my patients. The thought was so alien that it bothered me greatly. Eventually, I reassured myself that I had probably glimpsed her name in the appointment book for Monday.

First thing on Monday morning, I checked the appointment book – nothing for Anna! Later in the morning, there she was in the waiting room for a walk-in appointment. But why? Anna had been flying a Cessna home from the Tablelands on Sunday morning when the engine failed. She made an emergency landing in a cane field, while I was in my garden. She was a bundle of nerves, hadn't slept all night and needed some sedatives to calm her down. On the Sunday morning, I had received a mind-to-mind communication from her.

Two years ago, my house was on the market and was very slow to sell so I became well acquainted with the real estate agent. My gardening again was interrupted by thoughts of him, three times over a couple of hours.

I thought, "Why do I keep thinking about Mick, for heaven's sake?" I discovered the reason when I checked my phone later; he had sent three text messages while I was out of ear shot in my very large garden. I was receiving a thought message from a person who was trying to contact me.

Broadcasting telepathic thoughts was something I had wondered about over the years. I occasionally thought of a patient I hadn't seen for a long time. Within a day or two the person would show up saying that they weren't sure why they had made the appointment, but felt they needed to see me. This was a little disturbing; was this telepathy or had we both independently remembered a previously agreed appointment for around that time?

In this next story, there is no doubt that I accidentally broadcast my thought to a friend.

My friend and I had dinner at her home and shared a bottle of wine. As we ate, we chatted about our families, friends and eventually about ESP. She was disappointed that she hadn't ever had one of these interesting and fascinating experiences.

The next morning, I slept in and was running late for work. Slightly hungover, I struggled to get my soft, wobbly contact lenses to attach themselves to my eyeballs. First, I put them in the wrong eyes, then one of them turned itself inside out, then they both had bits of grit caught underneath. By the fifth attempt I was fuming, cursing and yelling inside my head, "These blasted contact lenses! I must ring the optometrist and make an appointment." Within two minutes, the phone rang. "Oh great," I thought. "Now I'm going to be even later for work!" When I picked up, my friend from the night before was on the line. She sounded very bewildered and confused and said, "Is this the optometrist? I think I must make an appointment." I said, "No, it's me and you have just received a telepathic thought!"

My friend lives at least one kilometre away, at the bottom of a hill and around the corner from my street. I wasn't talking out loud, using the telephone or semaphore but she picked up this thought about my contact lenses almost word perfect. I wasn't trying to send her a thought message; she received my thought via an unknown mechanism contrary to normal scientific laws. This was true telepathy!

2

What are Esp and Telepathy?

*Extrasensory perception (ESP) – also called the sixth sense – is defined as reception of information not gained through the recognised physical senses but instead with the mind. Telepathy is one form of ESP. Second sight or clairvoyance is the **alleged** ability to foresee the future (also known as precognition) or see actions taking place elsewhere (called remote viewing).* [1]

Wikipedia has a very negative definition and commentary on telepathy. **Telepathy** (from the Greek *tele* meaning "distant" and *pathos* meaning "feeling, perception, passion, affliction or experience") is the **purported** second-hand transmission of information from one person's mind to another's, without using any known human sensory channels or physical interaction. A simpler definition is: "mind to mind communication with another living person."[2]

The controversy arises with the words ordinary people use to describe both receiving and sending a telepathic thought. Typically, they say the thought does not belong to them; it has come from outside their mind which did not "think" the thought. Psychologists and psychiatrists call this phenomenon "thought insertion." Similarly, if someone receives a thought or telepathic message and voices it, the sender believes their mind has "broadcast the thought" or the recipient has stolen the thought.

Both thought insertion and thought broadcasting are hallmark symptoms of schizophrenia and psychosis. In fact, schizophrenia cannot be diagnosed without these symptoms. The thoughts have been described as auditory hallucinations and the psychiatric fraternity has turned themselves inside out arguing over the mechanisms involved in producing these symptoms. It has been the custom to diagnose anyone having these thoughts as having schizoid thinking, even if they are clearly not schizophrenic. However, a symptom is not a disease. Not everyone who has a cough has tuberculosis or lung cancer and not everyone who receives a telepathic thought has schizophrenia.

However, this is not the only barrier that prevents the scientific community from accepting ESP. The nature of ESP does not lend itself to scientific research. It is rare and sporadic, and may occur only once or twice in a person's lifetime, but is considered by that person to be extraordinary and unforgettable. It cannot be switched on and off at will. It often occurs at times of trauma or great emotion. Parapsychologists have conducted experiments and collected stories of ESP for many years, but their research has never gained acceptance in the general scientific community. It has been criticised on the grounds that the experimental design has lacked rigour and adequate protection against cheating and fraud. The later experiments conducted in the early 1990s have been criticised for the statistical methods used. Even when subjects correctly guessed one of five different Zener cards with a frequency that was statistically significant, this was not accepted as proof that telepathy or clairvoyance exists.[1]

Parapsychology is the name given to the study of ESP. It is considered a pseudoscience because it does not follow recognised scientific methods and has no acceptable hypothesis. A hypothesis is an idea about how something works, based on evidence such as observations and measurement and data. Such an idea is tested over time and by others to validate it. William Harvey's hypothesis about the circulation of the blood in the 1600s wasn't proven until microscopes were invented and we could see the capillaries that connect the arteries with the veins, enabling the blood to be circulated around the body and back to the heart. Perhaps ESP is no different and in years to come, we will have developed the technology needed to prove its reality.

From my late teenage years onwards, I have had several puzzling and unusual experiences which I can only conclude were a result of extrasensory perception. As time passed, I collected and cherished these strange occurrences, becoming more and more curious about this sixth sense.

My friend who received my thought about the contact lenses and the optometrist was alarmed when I told her it was telepathy and said, "Couldn't it just be a coincidence?"

"How can it be?" I replied. "You called me within two minutes of the thought exploding in my head and you were word perfect."

3

Telepathy or Coincidence?

Is telepathy real or is it all "just a coincidence?"
To answer this question, let's start by exploring the meaning of the word "coincidence".

The definitions of "coincidence" are very telling. Here is a selection from three of the world's most eminent English dictionaries.

- *Collins Dictionary*: "A coincidence is when two or more similar or related events occur at the same time by chance without any planning."[1]

- *Cambridge Dictionary*: "An occasion when two or more things happen at the same time especially in a way that is unlikely or surprising."[2]

- *Oxford English Dictionary*: "A remarkable concurrence of events or circumstances without apparent causal connection!"[3]

The key words here are *chance, remarkable, unlikely, surprising* and *causal.*

My patients and my friend thought that their telepathic experiences were just a coincidence or chance.

How we measure the likelihood of chance and coincidence

Mathematicians, scientists and biologists have ways and means of assigning a number to the likelihood of chance or coincidences being the cause of their experimental results.

They call this the **probability** – "p" – of an event occurring by chance. Gamblers talk about the same thing but call it the **odds**.

If a gambler is betting on the toss of a coin, the chance of throwing a head is one in two. On the second throw the chance of a head is also one in two, and so on for the third, fourth, thirtieth and five hundredth throw. The odds are calculated by multiplying together the chance for each throw – so the chance of throwing two heads in a row is one in four. The odds **against** throwing two heads in a row are three to one. Three heads in a row are one in eight and four heads are one in sixteen (written 1/16), and so on until infinity. As the number of heads in a row increases, so do the odds against this happening. The odds against four heads in a row are now fifteen to one against this possibility. Any gambler would know there was something funny going on if fifty heads in a row showed up. The chance of twelve heads in a row is 1/8192 and thirteen heads in a row is 1/16384. The gambler should have stopped using that coin long before twelve heads showed up because it is highly unlikely that this is occurring by chance; the coin has been weighted to land on the tail side. The question is: where do you draw the line?

Biologists draw the line at 1/20. They calculate the probability "p" by dividing one by twenty, to give a decimal number of 0.05 equivalent to five percent, meaning only five out of 100 times would the result occur by chance. The odds against it being chance are therefore ninety-five to five. Any results equal to or less than five out of 100 is not accidental, whereas any result greater than five out of 100 has occurred by chance. If we calculate the probability of twelve heads in a row, we come up with chance of 1/8192, and p = 0.0001 or one in 10,000. This means that this outcome is very, very unlikely to have occurred by chance, and our gambler will lose all his money if he continues.

Let's get back to my friend who phoned me for an optometrist appointment. What are the odds and probability that this was a chance event? Let's work it out.

First, we will consider the timing. She called within two minutes of the origin of the thought. Let's say she called within five minutes - to overcome any error in the timing of the call - and assume that there are fourteen social hours in the day to phone – say from 7am to 9pm. This would mean there are twelve five-minute time slots in an hour and fourteen hours available, giving a total of 168 five-minute time slots in a day. The chance of both of us thinking of making an appointment to see the optometrist within the same five-minute time slot is therefore 1/168 multiplied by 1/168 which is 1/28224, or a probability of 3.5/100,000 that this has occurred by chance.

Now let's consider the words. My thought was, "I **must phone** the **optometrist** for an **appointment**." Her question was, "Is that the **optometrist**, I think I have to make an **appointment**?" On the conservative side, we can say that she received two words – optometrist and appointment. It is reasonable to say that she also received the words "must phone", given that she phoned within a minute or two.

The word "appointment" is a noun. Each adult native English speaker has a minimum of 20,000 words in their vocabulary and of these approximately 3000 are nouns.[4] Word use has been analysed and the word "appointment" turned up three times in the top 1524 nouns, meaning it would be chosen once from 508 commonly used words.[5] "Optometrist" on the other hand wasn't listed in the top group at all, giving it a rating of 1/3000 nouns. However, some words go together in a sentence and some don't, for example "appointment" with elephant, ocean, or railway? "Appointment" fits with nouns describing a person's occupation, time and locality – such as boss, today or office. There were 170 out of 675 skilled occupations listed on the government website which would fit with "appointment".[6] So, a conservative estimate of companion words to be mentioned in the same breath would be about one in 300. This was the value I used for the word "optometrist.".

Mathematicians use complicated formulas to estimate the likelihood of the same thing occurring by chance, and you will find the formula I used and the calculations in the appendix (see Section 3).

The chances of my friend and I both choosing the word "appointment" were $p = 0.000004$, or four in one million. The chances of both choosing "optometrist" were $p = 0.000009$, or eleven in one million. The chance of phoning in the same five-minute time slot was thirty-five in one million.

It is permissible to combine these figures by multiplying them together, so the chance of this coincidence happening is one in one quadrillion. This unbelievable number translates to one in one thousand trillion.

My calculator couldn't compute the probability – the decimal number for p – it just read zero.

This means that the possibility of my friend knowing my thought by coincidence or chance is ZERO! A "coincidence" does not pass the pub test as the cause of our communication. Rather, telepathy is the cause of this *remarkable, unlikely* and *surprising* event, and this thought was communicated without using known sensory channels or physical interaction.

At this point, I asked myself a simple question: "How many people experience ESP and telepathy?" So, I decided to survey the patients attending my general practice in Cairns.

4

The Survey

The wet season in Cairns usually starts in mid-January and lasts through until mid- April. It is too early and warm for coughs and colds – the tourists stay away – so it is not a busy time in general practice. This is a perfect time to conduct a survey.

In March 2016, I started by asking each patient if they would be willing to take part in a survey about ESP. This question was almost always followed by a long silence and a searching stare. If they agreed, I then asked them three simple questions. Just under 170 patients were eligible. Of those, 148 patients answered all three questions, from a total of 500 adults who consulted me over that year.

Surveys of people measure all sorts of things from their size and diet preferences, their beliefs in religion and politics, to what type of dishwashing detergent is the best. Often, they confirm what we already know – such as most Ethiopians are very tall and thin compared to other humans – and nobody believes the results of surveys on dishwashing detergent.

When conducting a survey, it is very important that it is unbiased by the surveyor's and participants' beliefs and wishes, and that the survey sample mirrors the population from which it is drawn. If only Ethiopians were selected for a survey intending to measure the height and weight of people worldwide, we would mistakenly conclude that all humans were very tall and thin. Similarly, if I had asked my patients first about their beliefs in ESP and then only included believers, the results would have been as reliable as those about dishwashing detergent.

On the very first day, I was overwhelmed with stories from almost every second patient. I ended up running two hours late, which would never do! Subsequently, I looked at the appointments each morning and if the bookings weren't too heavy, I would choose that day as an ESP survey day. Every patient who attended that day was included in the survey, so I was able to collect a sample that represented my patient population. It took a very long time; the survey was spread over a year from March 2016 to March 2017. The questions I asked are included below. (The fine print of the survey is in the Appendix – see sections 1 and 2.)

The ESP Survey Question

Have you or any close member of your family ever had any kind of ESP experience?

This question was straight forward, designed to prompt a yes or no response. Very few patients needed an explanation of "ESP". It was deliberately non-specific and open-ended, so as not to influence them in any way. Doctors ask screening questions like this every day when we are taking a history of a patient's illness; for example, have you had any joint pain? If yes, the doctor and patient then tease out the details of which joint, when it hurts, what makes it worse, et cetera. If no, the doctor moves on to another question.

Seventy-two of my patients (46.2%) had at least one ESP experience themselves and another twenty had witnessed or been told of an experience by a close relative or friend, which indicated 59% in total had personal knowledge of ESP from a trusted source. Are you surprised? I was! I had not expected such a large proportion of my patients to answer "Yes" to this question.

When they elaborated on their positive response, patients were excited to share. The words came out in a great rush and tumble, often accompanied by lots of explanations about why the experience was so extraordinary, so unlikely, and defied common sense and logic. They usually placed the punchline first, so it took a lot of discussion to untangle the sequence of events – hence the need for plenty of time.

Who were these patients? - Just ordinary people coming to see their GP!

The sample included people from all walks of life – university students, homemakers, tractor drivers, farmers, sales assistants, administration workers, nurses, teachers, academics, the unemployed and invalided, healthy retirees, volunteers, artists, engineers, workers in tourism, small business owners, et cetera. The sample was a true representation of the patients attending my practice; however it was not representative of the Cairns nor the Australian population. My patients were older than the average Cairns resident, and women outnumbered men by more than two to one. The average age of the women was sixty-three and the men was sixty years.

According to the survey, men and women were equally likely to report a personal ESP experience and that of a close friend or relative (see Appendix/Section 1/Survey Statistics/Composition by Gender). Many people had more than one ESP experience and I collected a total of 159 stories from those with a personal experience. Those who had witnessed a friend's or relative's experience supplied another forty-four stories, making a grand total of over two hundred experiences. These included telepathy, clairvoyance, premonitions, visualisations of living people, visions of dead or dying people, sensing ghosts, hauntings, poltergeist activity, out-of-body experiences, helpful coincidences, sharing of pain with another person, and other strange things (see Appendix/Section 2 – the Experiences).

In recent years, there has been a disturbing shift in focus from the study of ESP itself, to the study of people who believe in ESP and other phenomena, such as magic, superstition, everyday religious beliefs, witchcraft, palmistry, UFOs, the Loch Ness Monster and reincarnation.

There have been failed attempts to link "belief in ESP" with low social standing, race, low intelligence, poor education, and various personality traits and styles of thinking.[1] In one study, belief in ESP was correlated with tests for magical ideation and proneness to psychosis, resulting from abusive childhood trauma.[2] Other research linked "intuitive thinking" to belief in ESP.[3] Others went further to suggest that believers have emotional – rather than analytical and rational – thinking and are suffering from widespread delusions held by the general population.[4]

However, some research showed that people who believed in ESP were those who had experienced ESP themselves and the more experiences they had, the more firmly they believed. In other words, "seeing is believing."[5] Believers were more likely to be women, students of natural or biological sciences, intelligent, creative, needing stimulation and variety, prone to fantasising, fearless of social ridicule, with good self-esteem and low anxiety.[3]

My survey did not ask about belief in ESP; it was purely an attempt to discover the prevalence of ESP experiences in the community. Prevalence (in medicine) is the proportion of people who suffer from an illness over a lifetime. It is described on a scale ranging from very common (more than 10%) to very rare (less than 0.1%).[6] Because my sample was older than the general population of Cairns and Australia, maybe there was more time for these patients to accumulate

ESP experiences. However, 47% is a very high percentage for personal experience of ESP; in my sample, it means that it is very common.

The second myth about ESP that needs to be analysed is that ESP experiences are linked to schizophrenia and/or are indicative of mental illness. Until recently, Wikipedia directed enquiries about telepathy to auditory hallucinations and mental illness, including schizophrenia and epilepsy. I explored the important question of whether ESP is related to mental illness by separating my patients into two groups: those who had been given a prescription for medication to treat a psychiatric or neurological illness; and those who had not been given such a prescription. Twenty-five percent of my patients were being treated with such medication during the survey period. This percentage is in the same ballpark figure as that for the Australian population, which is twenty percent.

Analysis of the results showed that the patients with neurological or mental illness had no more and no less personal experience of ESP than those without such medications (see Table 3 in Appendix/Section 1/Survey Statistics).

To be absolutely certain, the patients with the most severe psychiatric or neurological illness were separated from those with milder forms. This group suffered from severe depression, schizophrenia, bipolar disease, epilepsy, Parkinson's disease and other less common disorders. Twenty-five individuals were identified and compared with the group with no psychoactive prescriptions. Again, there was no difference between the two groups (see Table 4 in Appendix/Section 1/Survey Statistics).

One quarter of the people who had a personal ESP experience identified another family member as also having one or more of these experiences. Eight respondents declared their families were "psychic", with three or more members spanning three generations having ESP experiences. All these families kept their ability secret and never discussed it with outsiders.

The three important findings from the first question of the survey are:
- A personal ESP experience at least once in a person's lifetime is common, occurring in just under 50% of people in a general practice population in Cairns.
- Equal proportions of men and women have had at least one personal ESP experience.
- A personal ESP experience is not restricted to people with mental illness or neurological disease and does not occur more often in this group of patients.

Findings of other surveys on ESP

Other surveys have been conducted in the USA in the 1970s, and by Gallup's Multinational Study of Human Values in European countries in 1984. This included three paranormal questions. The questions related to the experiences of telepathy, clairvoyance and contact with the dead. The respondents were asked to select one of three answers for each question: Yes, No or Don't Know. However, the questions were not in the same language or phrased in the same way and not all participants answered all three questions. This means my results of overall prevalence of ESP cannot be accurately compared with surveys done by others.

Erlendur Haraldsson, a professor of psychology at the University of Iceland in Reykjavík, compiled the results of the Gallup survey with others done previously and research of his own in Iceland and Great Britain, and published the results in 1985 (see Table 10 in the Appendix/Section 2). In Western Europe, the most frequently reported experience was telepathy at 32%, followed by contact with the dead at 23% and 20% for clairvoyance. The prevalence of a personal experience could not be calculated from these surveys as the questions were asked separately and as stated above, not all questions were answered by each of the respondents. Therefore, the answers could not be simply added together, as a single individual may have had more than one type of experience.

The most interesting finding was that the results differed from country to country. Italy and France had the highest positive responses for all three questions – telepathy, clairvoyance and contact with the dead – while Denmark, Norway and Sweden had the lowest percentage of positive responses.[7] The reasons for the differences from country to country are not clear – it could be geographical, cultural or genetic.

In 2009, a survey conducted by Marilyn Castro and colleagues of the sociology department of Leeds Metropolitan University, England, used face-to-face interviews to ask five questions about telepathy, clairvoyance, precognition, after death communication and mystical experiences. This survey found 37% of the sample gave a positive answer to at least one of the ESP questions, and women were more likely than men to give a positive response to all five of the questions asked. People in the south-west of England reported the greatest number of personal experiences. This geographical difference could not be explained.[8]

The British response rate of 37% of those surveyed with a personal experience was lower than the result of 46% for the Cairns sample. However, this result in Britain reinforces the finding that extrasensory perception is a very common experience. The most common experience was precognition, or seeing the future, in 24% of the sample. This survey was representative of all people living in Britain, unlike the Cairns sample which is not representative of the Australian population.

Many years have passed since the European surveys. However, the types of experiences in contemporary Cairns are the same as they were in the surveys reported by Professor Haraldsson and those in England.

In the words of Andrew Greeley – one of the pioneers of parapsychology research who surveyed the US population in 1975 – "The paranormal is normal. Psychic and mystic experiences are frequent even in modern urban industrial society."[8]

The Telepathy Questions

The ESP question in my survey included telepathy as one of the categories of response. The survey also undertook a separate quantitative study of telepathy as a stand-alone experience. The questions I asked patients were:
- *When the telephone rings, do you ever know who is on the phone before you answer or look at the number on the screen?*
- *When you phone a person that you haven't contacted for a long time, do they ever say, "Oh, I was just thinking of you"?*

About five years ago, my landline telephone rang and I immediately thought, "That's Shirley on the phone" – a friend I had lost touch with for almost ten years. I was right – and I was dumbfounded!

Therefore, it seemed appropriate to open the telepathy section of my survey with the question, "Do you ever know who is on the phone before you answer?" The second telephone question was suggested to me by a patient; more than one friend had said "I was just thinking of you," when she phoned them after a long lapse of time.

These questions are like the two sides of a coin – both are an indication of telepathy, but the first question is about the reception of the message and the second is the sending of the message.

The results of my survey's telephone questions were very rewarding and quantified not only the prevalence of telepathy, but unexpectedly provided clues about its nature.

One hundred and fifteen patients (77.7%) responded "Yes" to at least one of the telephone questions. Men and women were equally able to receive or transmit telepathic thoughts. Eighty-one percent of women answered "Yes" to one of the telephone questions and 69% of men. This difference in ability was not significant by statistical analysis (see Appendix/Section 1/Survey Statistics/Table 5).

Patients volunteered the identity of the person they communicated with telepathically - more often an old school friend or workmate, than a relative, but someone they had known for many years. More than half of the telepaths communicated with two-to-three people but only two persons in the sample could identify many callers.

Analysis of the data identified 62% of the telepaths as receivers – they knew who was on the phone before they answered. Many people elaborated on this question, saying they would think of a friend who then called them within minutes. A few said their friend called within a day or two. The longest delay was one week – an old work mate had contacted the patient via LinkedIn, after a lapse of twenty years.

Senders were defined as those who called a person who then said, "Oh, I was just thinking about you," and these made up 80% of the telepaths. Some of the patients could both send and receive telepathic messages. This group was also more likely to have had a personal ESP experience (see Appendix/Section 1/Survey Statistics/Table 7), and they were also more likely to communicate telepathically with more than one person (see Appendix/Section 1/Survey Statistics/Table 8).

One very unexpected finding was that fourteen (11%) of the telepaths – a significant number – described **simultaneous**

phone calls with another person. What exactly do I mean by a "simultaneous phone call"? The following anecdote explains this well.

My husband was at work, having a coffee in his office, when he thought of phoning his daughter. He had an old-fashioned landline, so he picked up the receiver. As he put it to his ear, he heard her voice say, "Hi Dad," before the phone had rung or he had time to dial her number.

The patients from this cohort described similar events. The phone would have an incoming call from their friend in the same instant that they thought of the friend; the phone would be engaged because they were trying to call each other at the same time; a text message would arrive from their friend while they were dialling the number. In some cases, the same situation occurred with emailing or Facebook messaging.

Of the survey sample, eighteen patients identified themselves as having telepathic experiences in response to the ESP question. They reported "reading" another person's mind when they were close to them in the same room or when they touched the person. The information could be in the form of thoughts or of feeling the person's sensations. These people all answered "Yes" to one of the telephone questions and six answered "Yes" to both, indicating they were both senders and receivers. There was one exception – a woman who communicated telepathically with her husband over long distances but who answered negatively to the telephone questions.

The telephone has previously been used as a tool for paranormal research. In 2015, Rupert Sheldrake and volunteers conducted an experiment that was partly funded by the Institute of Noetic Sciences and the rest by private sources. Participants provided the researchers with the first names and telephone numbers of three people. The phone numbers were linked to an automated system, which then randomly selected one of the three people to phone the volunteer. The volunteer guessed which of the three was phoning and entered an identifying digit into the keypad before the call was put through. Mathematically, a correct guess should have occurred one third of the time. However, about 41% of the guesses were correct, and statistical analysis validated the use of the telephone to identify telepathic communication.[9] This experiment quantified the number of received telepathic messages. However, it did not and could not measure the prevalence of telepathic communication in a general population.

There are five important findings from the telephone questions asked in my survey:

- 77% of my surveyed patients had some telepathic ability.
- People who could both send and receive thought messages communicated telepathically with more than one person and were more likely to have had a personal ESP experience.
- Some people made simultaneous calls with another.
- 12% of the patients could gain information telepathically from another person's mind when they were physically close to that person.

The results differed greatly from those reported in the surveys by Professor Haraldsson. The highest percentage of people with telepathic ability lived in the USA (58%), followed by those in Italy, France, West Germany, Finland and Iceland, with about 35% in each.[8] Overall, the British survey showed that only 11.5% of participants had experienced telepathy. This equates with the eighteen people (12.2%) who reported a telepathic experience in response to the ESP question in my survey.

The other surveys did not use the telephone questions, which could be part of the reason there is such a huge discrepancy.

There are two possible reasons why the telephone questions yielded these results. Firstly, there is the interpretation of "Oh, I was just thinking of you" as a coincidence, even though the caller may not have phoned the recipient for many months and the timing of the call was not linked to a special event such as a birthday or anniversary. Most of the patients did not recognise this as a telepathic communication. However, for the receiver to be aware of the identity of the other person, information must have been transmitted other than by the normal senses.

Secondly, the telephone imparts an immediacy and importance to a *surprising*, *remarkable* and *unlikely* event, which is hence recognised and remembered as something very odd, rather than as a coincidence.

Whatever name is given to these experiences – ESP, psychic, paranormal or supernatural – they are common in our society and encompass telepathy or mind-to-mind communication, precognition, clairvoyance, telekinesis, sensing of ghosts and others.

The survey yielded results that quantified the prevalence of extrasensory perception and telepathy in a small patient cluster. We will now explore the details of these telepathic messages, which provided further information about the nature of telepathy.

5

Telepathic Experiences

I will begin with my experiences and continue with those of my patients. Each of these accounts adds to the knowledge of telepathy with soft – or non-quantifiable – data.

Almost a decade after leaving school, I returned to university to study medicine. To attend classes, I used to walk and catch the ferry across the river to the University of Queensland. Once a week, I had a free afternoon. One afternoon, as I walked through the gate of my home, a thought came to me.

Ian is coming to see me today.

Ian and I had been very close when we were university students previously, but life had taken us in different directions. Although we kept in touch, I hadn't heard from him for months and hadn't seen him for at least three years.

I didn't think any more of it as I settled down at home to study. Around nine o'clock that night, the gate creaked. There was a knock on the door and there he was. I was overjoyed to see him, but I completely freaked him out when I said, "I knew you were coming to see me today."

Many years later, I had married and was living in Cairns, working in my own practice as a GP. I saw a patient every fifteen minutes and disciplined my mind so that each patient had my full attention. Consequently, it was difficult to relax after a day at work. If I played patience on my computer for an hour or so before bed, I could get to sleep more easily. One night, I was happily playing when the thought came: *Julie is having her baby tonight.*

Julie was my friend and though she looked very pregnant, she wasn't due for another six weeks. I thought, "That can't be right. Why did I think that?" The next day, I discovered she'd had triplets the night before by emergency caesarean. She had kept the fact that she was carrying triplets a secret!

Another incident was very strange. Again, I was sitting at my desk playing patience online when the thought arrived: *Bill G is dead.*

This time I was completely flabbergasted. Where had this thought come from? Bill G was a medical colleague, but I didn't know him well and had little personal contact with him. I found out later that he had died unexpectedly that evening from recently diagnosed cancer. I had no idea that he had been ill.

The following stories are those of my patients. There were many of these – often very similar to each other. I have selected a few to illustrate some of the different circumstances in which telepathic messaging occurred.

Common tales of telepathy

Bernard belongs to a Christian community. When he called a friend from his church whom he had not seen or spoken with for almost nine months, she said, "Oh, I was just thinking of you and that it was time that you rang me."

> *Leanne is a sales assistant in a delicatessen. Once, while on the way to work, she thought, "I haven't seen my friend Ainslie for quite a while." That day, Anslie came into the shop with some plants for Leanne's garden.*

Simultaneous phone calls

Fourteen of the surveyed patients had made simultaneous calls with one other special person. These are some of their experiences.

> Ken has an old friend he typically calls only once a year. The last time he called, his friend said, "Bloody hell – I was just this minute talking to the missus about you." On another occasion, Ken said to his wife, "We haven't heard from Alison for a long time." Almost immediately, a Facebook message from Alison – who lived in Broome, Western Australia – appeared on his mobile phone, saying that her mother had died in England.

Telepathy existed before the telephone was invented but it would have been impossible to prove that two people were thinking of each other simultaneously, making it easier to explain this away as a coincidence. Some of my older patients remembered incidents from their early years, when they would think of another person and then run into them shortly afterwards. They would say, "Speak of the devil – I was just thinking about you!"

Telepathy and mindreading

Kayla has telepathic communication when two or three different people call her on the phone. She will say to her husband, "Can you get the phone please? It's so and so." And she's always right! She is also able to "read" another's mind when physically close to them. In the past, she supervised carers of children with disabilities. When the carers had questions, she knew what they were about to ask and would pre-empt them saying, "Yes, you can take Johnny to the swimming pool but be back by three o'clock." When they asked how she knew what they wanted to know, she couldn't explain. The ability to read their minds extended to all the carers.

> Kate had a similar rapport with a colleague she had worked with for eight years in the cosmetic industry. When she entered his office with a query, he would often tell her the answer before she had asked the question. She told me, "We seemed to be inside each other's heads!"

> Heather and her husband, who live separately, often think the same thought at the same time. One day, she was browsing the internet, for real estate in Noosa for no special reason. Her husband phoned and said, "I want you to go on the internet and have a look at this property in Noosa." When he described the property and gave her the address, she realised that she was already looking at the very same house. Interestingly, this patient said "No" to both the telephone questions, even though she was clearly having mind-to-mind communication with her husband.

Telepathic warning/bad news

Some patients told of knowing when a loved one was in danger.

> Sibyll's son was studying at university and living in the family home. One day, she was hanging out the washing in the backyard, when she noticed thunder, lightning and a terrifying, huge dark storm cloud approaching. She found herself hanging onto the clothesline, praying to God as hard as she could, "Please take me and not him – he's too young." At that time her son had lost control of his car and crashed into a telephone pole miles away from home. The pole snapped in half and crushed his car. His arm was broken; his glasses were

> smashed but thankfully he survived.

> As a young woman, Marilyn shared a flat in London with a friend. One Sunday morning, they were having breakfast when her flatmate suddenly sat bolt upright and exclaimed, "Something just happened to my twin sister in New York!" Her sister had been injured in a motor vehicle accident.

Telepathy/good news

Not all telepathic messaging is the harbinger of bad news.

> Patricia rang her mother in Perth every week. She was surprised one day when – without even saying hello – her mother blurted, "You're pregnant!" This was true! The same happened again three years later, when Patricia was pregnant with her second son.

> James, his ex-wife and daughter all know what each other are thinking, although they live separately. One day, James put on a roast for dinner and his daughter phoned her mother and said, "Dad is cooking a pickled pork roast." This family communicates trivial information every day, telepathically.

Sharing sensation or pain telepathically

Not all telepathic communication is of thoughts or knowledge. Some people also share pain or feelings telepathically.

> Two of my patients – Susan and Deidre – were nurses and they both told me that when they touched a patient, they felt that person's symptoms, such as nausea

or pain. They both believed that this helped them with the patient's care.

Laura was in a sewing class at school where the girls were learning to use a treadle sewing machine. The needle went through the fingernail of one of the girls. Her twin sister, who was in another part of the school, felt something wrong with her own hand and came running into the room to find out what had happened.

Helen's father was working at a road camp outside of Childers. One night, he developed such severe abdominal pains that he couldn't sleep. Unbeknownst to him, his wife was in labour at the Nambour Hospital, where she delivered Helen shortly before morning. Sixty or so years ago, there was no telephone communication available with such a camp. The doctor who brought Helen into the world offered to take a message to the road camp and found her father sitting up in bed, having a cup of tea and fully recovered from his night-time ordeal of "labour pain".

Societies that use telepathy as a means of communication

In 2011, a friend from Cairns travelled around the Calvados Archipelago of the Louisiade Islands off the eastern tip of Papua New Guinea. My friend, his son and three others travelled in a sailau – an ocean-going canoe with a large sail and an outrigger – which required a crew of ten locals. They camped on the beaches of some very small, isolated islands. The villages have no electricity and there are no mobile phones, internet or radio. Juda, the owner of the sailau, was adept at "magic", which is considered by the islanders to be very powerful. My friend enquired how Juda communicated with those on other islands.

He replied, "If I travel tomorrow, I send a message in mind. My friend sees my face in his dream, then he knows to get ready because I come. When I arrive, they are ready with a feast and ceremony."

David Unaipon was a remarkable man – an Aboriginal of the Ngarrinyeri people, the son of a preacher, a student at the Point McLeay Mission School in South Australia and in turn a bootmaker, bookkeeper, preacher, author and inventor credited for the invention of mechanical shears amongst other things. Although he suffered discrimination throughout his adult life, he is recognised today with honour and graces the back of the Australian $50 note. The following article, entitled "Aboriginal Telepathy – Remarkable Explanation," was published in a Victorian newspaper The Argus, dated Thursday 6 August 1931:

David Unaipon, a full-blooded member of the Ngarrinyeri [sic] tribe of Aborigines, aroused keen interest among the members of the Victorian Institute of Advertising at lunch yesterday by describing to them the method adopted by the aborigines when sending messages over long or short distances. "When an Aborigine wishes to appeal for help or to send any other message to another member of his tribe, he first attracts attention by a smoke signal. The man who sees the smoke signal then strives to do a very difficult thing – to clear his mind of every thought and so to become fully receptive to messages sent to him. The man who made the smoke signal concentrates his thoughts on the desired message and soon it is received and retransmitted to the rest of the tribe. At night when a smoke signal would not be seen, the Aborigine waits until the person he wants to communicate with will most likely have lost consciousness in sleep. His subconscious mind is then fully awake, and it will receive the message.

This is almost identical to the explanation given by Juda of the Louisiades.

All these patients who shared their experiences with me considered them to be remarkable and out of the ordinary! Fascinating as they are, what do these stories tell us about telepathy? Can we extract from them any common laws about the nature of telepathy?

6

The Nature of Telepathy

Why is this chapter important?
　　The history of science and the scientific method are based on the collection of evidence about naturally occurring phenomena – such as sunlight – so that the phenomena can be understood. The laws explaining the nature of a phenomenon can be derived and expressed in mathematical terms, then harnessed for our benefit in our modern world. If the study of telepathy and ESP is ever to progress from a "pseudoscience" to true science, then we must measure and describe the properties of telepathic or "thought" messages.

Let's consider sunlight as an example. Initially, we knew that sunlight supplied light and heat. It passed through the glass in windows but was blocked by a brick wall. It caused shadows. It was reflected from the surface of smooth water, but the image was upside-down. We only discovered that it was refracted into the colours of the rainbow when prisms were invented, and much later that it could be used to produce electricity. Now, we know the mathematical equations; we know how much energy is transmitted by light. We use this knowledge about light in laser technology for eye surgery, transmitting information to our TV screens and pinpointing targets for weapons.

Telepathy is at the beginning of this scientific process. The telephone questions have shed some light on the physical properties of a "thought" message, and the stories of patients' experiences have contributed details about its nature.

What can we discern from these stories of telepathy so far?

- The nature of the telepathic message is vague. Most of the patients who answered "Yes" to the telephone questions were only able to identify the caller – they had no idea what the person was going to say or what news they had to impart.

- Most people can only receive or send a telepathic message to one or two people – usually a relative or close friend.

- Telepathic messages are few and far between. Over a period of fifty years, I have received less than a dozen messages and broadcast only one that I am certain of. They are not usually an everyday occurrence unless you belong to a family like James', where they communicate telepathically about dinner menus.

- It seems that each of the messages I received was sent by a person who was in a highly emotional state: my boyfriend coming to see me after a long separation; my patient terrified by her aeroplane's engine failure; my friend having triplets and again, when her husband's best friend Bill died. I was extremely angry, when I accidentally sent my friend the message about my contact lenses and the optometrist.

- The messages I have received only seem to appear when my mind is quiet, when I am relaxed and not actively thinking. I was walking home from university and received Ian's message; I was gardening and received Anna's; I was playing patience on the computer with the other two messages.

- The nature of the thought is special. It is out of left field – is alien or foreign – and does not originate in my own mind. It clearly matches the psychological description of thought insertion. Each time my immediate response has been "Where did that **thought** come from?" I do not **hear** the messages. It is more like **knowing** a fact – such as "tomorrow is Christmas day."

- The messages are not usually specific or detailed. I didn't know that Anna had an emergency landing in her aeroplane – only that she was coming to see me the following day. I didn't know that Julie was having triplets – only that she was having her baby that night. The message I sent to my friend about the optometrist was very detailed and she received it word perfect. This was very unusual in my limited experience.

- The message is more detailed and accurate if the sender and receiver are close to each other: Kayla and Kate both "knew" their co-workers' thoughts when they were in the same room; my friend received the message about the optometrist one kilometre away. At larger distances, as with Ken, only a thought of the sender seems to come through – not the detailed information.

- Telepathic messages travel very fast and are probably almost instantaneous. Using the telephone questions as an assessment tool allowed fourteen of my patients to recognise simultaneous messaging.

- Telepathic messages can travel over very long distances that stretch thousands of kilometres: my husband in Cairns having a simultaneous phone call with his daughter in Brisbane; Ken in Cairns receiving a simultaneous message from Alison in Broome, WA.

- It is possible to identify the sender of the message. There was no uncertainty that it was Anna and Ian who were coming to see me. My friend with the optometrist message did not ring her optometrist – she rang me! Subconsciously, she had identified me as the originator of the message. This means that an individual's thoughts must be identifiable.

- Telepathy is an ability that we may be able to fine tune, so it becomes more noticeable and accurate with practice. One of my receptionists was aware I was conducting this survey with my patients. Towards the end of the twelve-month research period, she told me that she was becoming increasingly aware of telepathic messages. She would think of a patient and within fifteen minutes or so, that person would ring to make an appointment. This was happening once or twice a day.
- Lastly, these thought messages were verifiable, true events.

What have we found?

Telepathy is the transmission of a thought message or of a bodily sensation, such as pain, from one person's mind to another.

The qualitative properties

- The message received can be vague or very specific.
- The sender is probably emotional – frightened or angry – and the receiver's mind needs to be quiet.
- The sender can be identified.
- The receiver recognises the thought as foreign.
- It is a rare event for most people.
- Telepathic communication usually takes place with only one or two close relatives or friends.
- We can become more aware of telepathic communication.

The quantitative properties

- Telepathic messages travel very fast and far and probably are almost instantaneous.
- In mathematical language, we can say that the amount of information transmitted by a thought message is inversely related to the distance between the two participating minds.

Observation and curiosity about "how things work" are at the beginning and heart of our research into any mystery in biology, physics, our world and our Universe.

We have only just started our journey into investigating ESP. We know from the surveys – mine and others – that many people have these experiences; they are very common. Belief in ESP is widespread but there is scepticism, doubt and confusion about whether it is "real" – which leads us to the next section of this book.

Can we make any sense of how telepathy might work?

Part II: Science of the Mind

How does Telepathy Work?

Many people are intrigued by the notion of telepathy and extrasensory perception. The minds of physicists, engineers and astronauts have been grappling with the problem of "how does it work" and have come up with theories related to time, the existence of multiple dimensions, the zero-point field and lately quantum theory. Religious and spiritual people look for understanding in prayer or the "world of spirits". On reading, these theories seem to be as tortuous and inexplicable as each other and none of them shed light on how we can read another's mind or feel another's pain.

I sought some help from a quantum physicist who listened patiently to my two signature descriptions of sending and receiving telepathic messages. She said, "I think the brain is acting as an antenna for electromagnetic radiation!" This idea has been explored in the past and has been "left on the shelf", but neuroscience and knowledge about the brain have blossomed exponentially in the last forty years.

In science, an idea about how something exists or works is called a hypothesis. It is important to have a hypothesis about telepathy because it allows experiments to be carried out to test the idea and see if it is feasible or not. Exploration of the idea that the brain is acting as an antenna and that telepathic messages are transmitted by electromagnetic radiation has been fascinating and has led me down many different paths, into areas I would never have expected.

Come with me on this journey and we will see how this might work.

7

Electromagnetic Radiation and our Senses

What is electromagnetic radiation?

On planet Earth, we are bathed in electromagnetic radiation everywhere and all the time. It originates in the Universe from our sun, the stars, quasars, black holes and on Earth by electrical currents moving through the air of the ionosphere. In the last two hundred years, we have learned how to produce electromagnetic radiation and we use it to communicate with each other by wireless, telephone, TV and computers.

Electromagnetic radiation is generated from a vibrating electrical source. This source creates an electric field that swings backwards and forwards from negative to positive, like a pendulum. The electric field in turn generates a magnetic field – its twin – which also swings backwards and forwards, but at right angles to the electric field. The two fields between them create a wave of energy – electromagnetic radiation – that also swings and travels at right angles to both the fields. If you imagine them as lying flat on a piece of paper, with the electric field pointing north–south and the magnetic field pointing east–west, then the electromagnetic wave will travel upwards and downwards into the air – perpendicular to the sheet of paper – to reach your eyes or your toes (see Figure 1 in Appendix/Section 4).

Electromagnetic radiation swings as it moves forward so traces a wave like pattern to form a cycle. Each cycle looks like the letter "S" lying on its side and is joined to its neighbours to form sine waves. Sine waves resemble the ripples on a pond made by a falling stone. The power or height of the "ripple" is called the amplitude; the distance between two "ripples" is called the wavelength; and the number of "ripples" or cycles in one second is called the frequency.

Electromagnetic radiation travels forward as a procession of sine waves. Scientists measure three properties of an electromagnetic wave to assess its power: the electric field, the magnetic field and the energy the wave possesses.

Electromagnetic radiation is the Universe's mobile energy. It zooms through space at the speed of light; it can travel vast distances; it passes through solid objects; it can be used to carry information. It travels in straight lines unless it is bent into a curve by gravity as it passes a star, and it can be reflected like light from a mirror. When it arrives at a destination or meets an atom along the way, it releases energy as quanta – teensy packets of energy. A quantum of light is called a photon. Electromagnetic radiation comes in different sizes and strengths. The electromagnetic spectrum starts with extremely low frequency (ELF) waves (very weak and up to 300 cycles per second) and progresses through radio waves, microwaves, infrared light, visible light, ultraviolet (UV) light, X-rays and finally gamma waves. These are the most powerful, with a frequency of trillions of cycles per second. They are produced by stars or by splitting the atom in nuclear explosions, and can penetrate our bodies, destroy our DNA and kill us.

Electromagnetic radiation and our normal senses

We can't "see" most of the electromagnetic spectrum with our human senses – only those frequencies that produce visible light. Absolutely everything we know or feel about our world travels from sensory organs to our brain as electrical signals. We "see" our world because photons from the electromagnetic radiation energise electrons in cells in the back of our eyes, which then emit small electrical signals. These signals travel to our "visual" brain, where they interact with neurones. Many such signals and neurones build an image which we "see.".

Other creatures have access to senses that we don't possess. Bees and insects use more of the electromagnetic spectrum than we do. They can "see" UV light which enables them to identify petals and flowers, and polarised light to aid navigation. Snakes can "see" infrared waves as heat, which helps them find prey in the dark. Sharks have an organ called the lateral line which allows them to "see" the electric field of marine animals. Bats and dolphins can sense high-pitched sound, which they use as echo location. Until the advent of modern science, we were unaware that living beings could use these types of energy.

Wikipedia defines telepathy as "the purported transmission of information between people without using known sensory channels or physical interaction." This begs the question: what are the known senses and how do they operate to allow people to communicate?

The known senses

Tradition allows us five senses – those of sight, hearing, touch, taste and smell. Over the years as our understanding of the human body has grown, more have been added. Our bodies are permeated with nerve endings that are specialised to detect changes in ourselves as well as our environment, such as those that monitor pressure in our joints and movement of our muscles, so that we know where our legs and arms are in space.

Our thoughts are transformed by our brains into speech, writing, facial expressions – which often give the clearest messages of all – as well as hand signals and body language. Thoughts and feelings transmitted in this way are open to interpretation, so their meanings are not always clear. However, regardless of how a thought is interpreted, it is obvious that information has been transmitted by one or a combination of these known ways.

So, now we come to the question: how can we share information by telepathy using unknown senses? Do we possess senses that haven't yet been discovered and/or is there a way of transmitting information without needing physical interaction? Could there be real mind-to-mind communication via electromagnetic radiation?

The next question is: "what is thought?"

This has exercised the minds of philosophers for many years. Is a thought real? We can't see, hear or smell our thoughts. How do we know they exist? Common sense would say that our thoughts become real with communication to others, and that our thoughts are real to ourselves as products of our mind. Our mind is also unable to be seen or heard or touched, and there are an equally large number of philosophers and scientists concerned with theories of mind and consciousness.

Our electrical bodies

We think of ourselves as flesh and blood, but in fact we are electrical creatures – as are all other living animals and plants.

If we could "see" the electrical activity in our bodies, we would know that we are electrical beings with a blinding, bedazzling concentration of electricity in our brain. Every sensory nerve in our body is ultimately connected to our brain and information is channelled along these nerves by intermittent electrical currents. Every heartbeat, every word, every movement is initiated by an electrical current in our nervous system. We generate an enormous amount of electricity daily – about 25 watts per second in our brain alone which has more than eighty billion neurones, each with a negative electrical charge at rest of minus 70 millivolts/millimetre. This sounds very small compared to 4 volts/metre produced by the battery of your mobile phone, but when a neurone's microscopic size is factored in, one neurone produces an electrostatic force equivalent to 14 million volts per metre. This is equivalent to four times the energy of lightning discharged by a storm cloud.

All electrical conductors produce an electric field. The switches in your house, the cords carrying electricity to your TV set, your phone charger and toaster all have an electric field which extends beyond the wires even when they are in sleep mode. Similarly, we too have an electric field, due to the flow of current through our nervous system. Our electric field extends outside our skin for a short distance. We could think of our flesh and blood as the necessary equipment to maintain our electrical energy.

How does this relate to thought and the transmission of thoughts between people? If thoughts were to be broadcast as an electromagnetic wave, then we would need to have an electric and a magnetic field interacting with each other to produce such a wave and our brain would need to have a mechanism to receive such a wave.

In the next chapter we are going to explore how our brain generates an electric field, brainwaves, and the structure and organisation of the computer that is our brain.

8

Neurones and Brainwaves

The brain is the most complex computer on our planet, built not from metal and silicon chips but from flesh and blood.

Neurones are the building blocks of the brain and nervous system. Each neurone is an electrical sensor, battery, switch and the wiring that carries an electric signal to another neurone, muscle or gland cell.

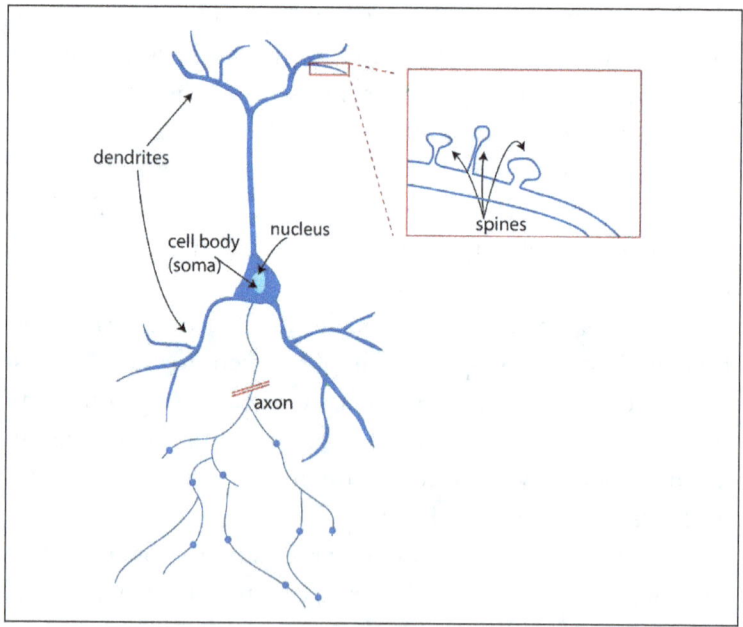

Figure 1: Neurone of the cerebral cortex
Image credit: The Queensland Brain Institute at the University of Queensland.

This is the basic design of a neurone. The dendrites (from the Ancient Greek meaning tree) sense the incoming electric signals. They have many tiny buds called spines and these have hundreds of synapses, which are the contact points between one neurone and another.

The body of the neurone has an area very sensitive to electric charge where the axon arises. The axon is the "wiring" which carries the electric signal from one neurone to the next. It is insulated by layers of cell membrane containing fat and cholesterol along its entire length, just like the plastic that insulates copper wires in your house. The electrical signal passes down the axon to terminal buds and their synapses. The signal then is transmitted across a tiny gap to receptors on the dendrites of another neurone.

The neurone is a small battery!

In a car battery, an electrical current is generated by electrons moving through copper wires. In the brain, it is the positively charged sodium and potassium ions passing through ion channels that create an electric current.

Every cell in our body contains and is surrounded by water. The brain and spinal cord float in a bath of water called cerebrospinal fluid, or CSF. This fluid is 99% water and contains glucose, a tiny amount of protein, and ions – atoms with an electric charge – including sodium, calcium, potassium, magnesium and chloride ions.

The neurone is enclosed by a waterproof membrane, made of a double layer of unsaturated fat and cholesterol, which is only five millionths of a metre thick. The membrane creates a barrier so that sodium ions collect along it on the outside of the neurone and potassium ions collect on the inside. The ions can pass from the outside to the inside of the neurone through channels which are opened or closed by chemical or electrical (voltage) gates. Sodium and potassium ions have the same positive electrical charge, but because there are more sodium ions on the outside than there are potassium ions on the inside, the inside of a resting neurone has a negative electric charge of minus 70 millivolts, compared to the outside.

When an electric signal arrives at the end of an axon, neurotransmitter chemicals are released into the CSF, filling the small gap at the synapse. They float across the gap and lock onto neuroreceptor proteins on the dendrite side. This opens the gates of the sodium channels by a chemical reaction and the sodium ions flow into the neurone, increasing the number of positively charged ions inside so that the voltage increases a little. When it reaches about minus 50 millivolts, voltage-gated ion channels suddenly spring open and huge numbers of sodium ions flood into the neurone, within just 0.2 milliseconds (one five-thousandth of a second).[1]

The neuron is now full to bursting with static electricity (about plus 40 millivolts) and a sensitive structure at the first part of the axon discharges the electricity, which travels at great speed down the axon to its terminal buds.[2]

The electric discharge is called firing, a spike or an action potential (AP). Each AP always has the same power and speed and lasts for about 1 millisecond (see Figure 2 below). For instance, the spike seen on an ECG recording of the heart, which triggers a heartbeat, is an AP.

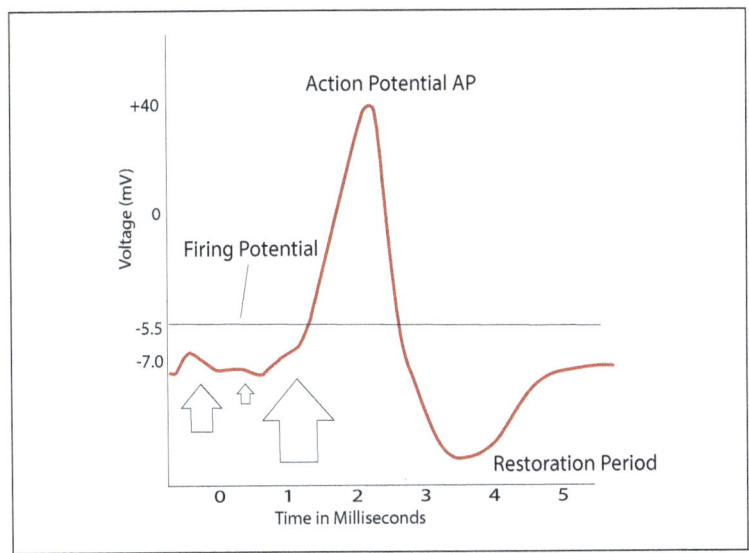

Figure 2: An Action Potential
The neurone must be reset after it fires, and this requires energy produced from glucose or fatty acids. The energy is used to pump out three sodium ions and pump in two potassium ions, so that the neurone can fire again. Twenty percent of the calories we eat every day is used to reset neurones in our brain.

Neurones are switches!

An AP is an all or nothing event. If the signal reaching the dendrites is not strong enough to allow sufficient sodium ions to pass into the neurone and reach the threshold, then the neurone won't fire. Some neurones use neurotransmitter chemicals that activate chloride channels. Chloride ions have a negative electric charge which makes the inside of the neurone more negative, so a bigger signal than usual is required to reach the firing threshold.

The neuron acts as an on/off switch by firing or not. The signal is relayed only if the neurone fires off an AP. In modern terms, the on/off states act like a binary code used in computing systems.[1]

So now we know how a single neurone works. But we have 85 billion of them — what are they all doing?

Neurones produce an electric field, sine waves and brainwaves.

Neurones with both positive and negative input have rhythmic pulses of ions that move like waves on the beach, washing in and out through the ion channels. This causes a swinging electric field, producing sine waves just like those discussed in the chapter on electromagnetic radiation. Different types of neurones produce different frequencies of sine waves, and each neurone has its own frequency in the range of one to about 140 cycles per second or hertz (Hz). Frequencies are organised into groups, and each group has a name: delta for 1–4 Hz; theta for 4–9 Hz; alpha for 10–12 Hz; beta for 13–30 Hz; and gamma for 30+ Hz.

The sine waves travel through the brain tissue and combine with each other — mixing, matching or even cancelling each other out — to form brainwaves. These are measured on the surface of the scalp and the printout is called an electroencephalograph (EEG).

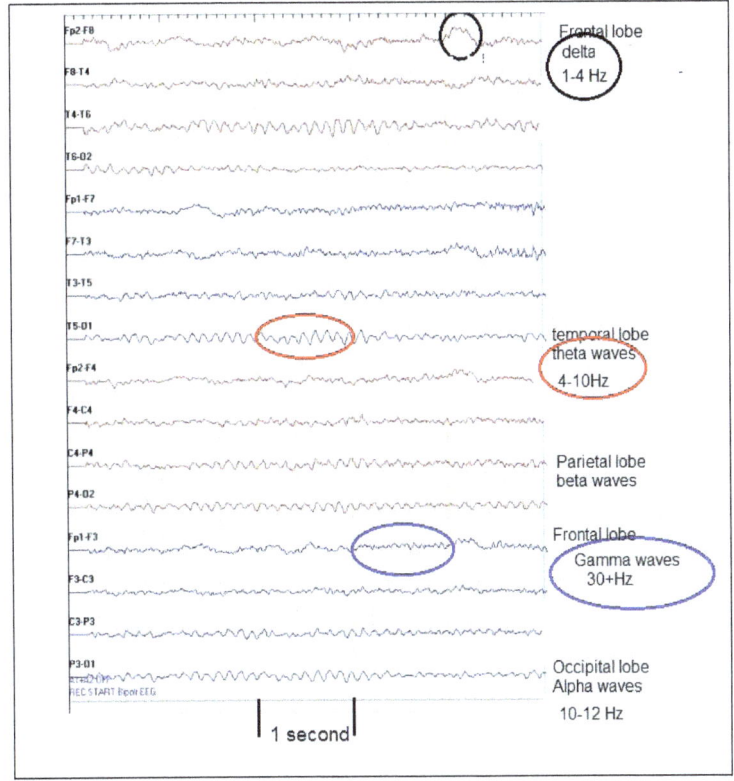

Figure 3: A typical scalp EEG recording
Each line on the EEG represents a different region of the brain; the redlines are from the right side of the brain and the blue lines are from the left. Like the ripples in the pond, the height of each wave is its power in volts and the number of waves in one second is its frequency measured in Hertz.
As you can see, the EEG is a mishmash of squiggles of differing shapes and sizes. The most vexing question has been: do the squiggles or brainwaves mean something or are they just gobbledygook?

Advances in computing have allowed us to analyse brainwaves in more detail.

Quantitative EEG (qEEG) uses computer software to separate the brainwaves into individual frequencies and measures the voltage at each frequency, which led to the discovery that the "visual" brain uses alpha waves.

Functional magnetic resonance imaging (fMRI) scans measure the oxygen levels in the different areas of the brain, enabling us to locate the areas of the brain that are active while remembering words or performing mental arithmetic, for example.

Microelectrodes placed into the brain tissue are used to measure the activity of single neurones. This is an invasive technique which could potentially damage the tissue or cause infection. As such, it is only used in laboratory animals or in epileptic patients undergoing neurosurgery to treat their disease. Microelectrode arrays can measure about 100 neurones simultaneously and record the patterns of firing.

So much for the neurones, electrical signals and brainwaves – how does this all function as a computer? We need to take a quick look at the organisation and structure of the brain.

9

The Human Brain

The brain is the central station of the nervous system. It receives information from the outside world as electrical signals, combines and coordinates this information, and produces outputs as thoughts and actions.

The human brain looks like two cauliflowers – joined together by a thick white stem – each of which is called a cerebral hemisphere. The top, bottom and sides of the hemispheres are wrapped in a greyish surface layer called the cerebral cortex or "grey matter", which is made up of the dendrites and bodies of millions of neurones. The "white matter", which lies beneath the grey matter, is made of the axons of the neurones, and is arranged in bundles that carry information from one part of the brain to another.

The cerebral cortex is folded into deep fissures and hills, so that its total area of two square metres fits inside the skull. Special relay centres are scattered through the older parts of the brain and spinal cord. These relay centres are made up of neurones which transmit the primary electrical signal coming, for example, from your toes via a synapse to a second neurone in the spinal cord or brain stem, which then relays the signal to the cerebral cortex. These are like the nodes of the national broadband network, which relay signals from fibre optic cable to the telephone wires entering your home and enable your friend at one address to message you at yours.

The electrical signals are not received higgledy-piggledy; they are carried and arranged geographically at the relay centres and at the cerebral cortex into GPS-like maps, so that a signal from your **left** toe arrives at the left toe-address in the **right** side of your brain!

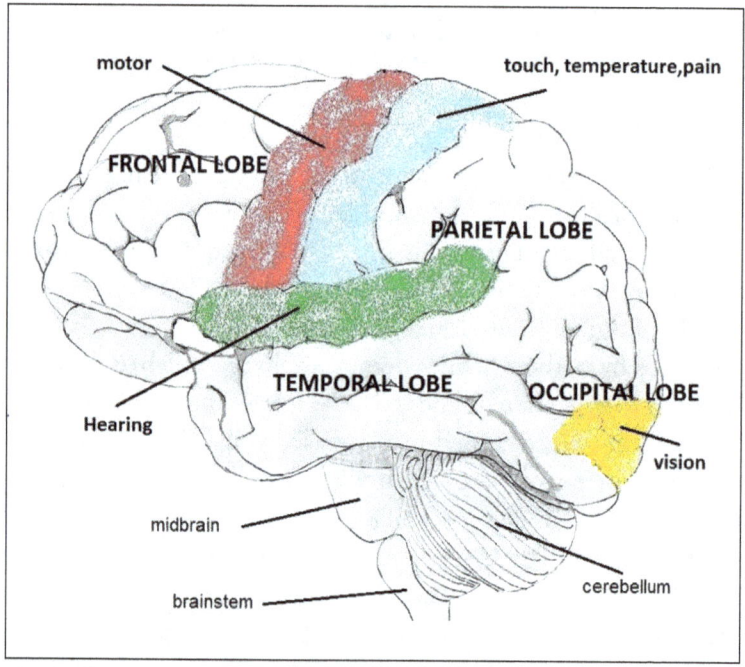

Figure 4: The human brain
The brain is organised into separate areas which have different functions. Each hemisphere is made up of four lobes and multiple regions. The sensing areas, **receiving** *signals from the outside world, are located in different lobes: vision in the occipital lobe; hearing in the temporal lobe; and skin sensors of touch, pain and temperature in the parietal lobe. The motor cortex in the frontal lobe* **sends** *electrical signals to the body. The white areas in Figure 4 are association areas where neurones collect signals from many senses, and combine, integrate and coordinate these signals to construct specific thoughts or actions, such as riding a bicycle.*

If you look back at Figure 3 in Chapter 8 (the EEG), you can see that the brainwaves vary in size and shape, and that different areas of the cortex produce differently sized waves. The top line recorded from the frontal cortex (memory and thought) looks quite different to the bottom line from the occipital lobe (vision). Each region has its own preferred frequency of sine waves. The frontal lobe uses combinations of delta, theta and gamma waves; the occipital lobe uses alpha and gamma; the motor and tactile sensory areas use beta; and the sense of hearing located in the temporal lobe uses gamma frequency.

In the next chapter we will discover how the neurone-generated brainwaves carry information from one area of the brain to another, and how thoughts appear in our mind.

10

Coding Thoughts

How does the brain receive useful information about the outside world through electrical signals and process this information into thoughts and actions?

Picture me sitting on my verandah reading a book, when a March fly bites me on the leg.

The pressure of the bite activates the ion channels of a few touch and pain neurones, which send off APs. These signals travel more than one metre to reach the first relay centre in my brainstem, and if they are too weak will end there. To reach the "leg address" in the sensory cortex of the parietal lobe, the signal is strengthened by the neurones firing off repeated APs or bursts of APs like a machine gun.[1]

Meanwhile I am oblivious to these signals, concentrating on my book.

The axons of the relay neurones branch out and send the same signal to many neurones in the sensory cortex, which are arranged parallel to each other in small clusters called ensembles further magniifying the signal.

Now my brain can detect **where** on my leg the fly is biting.

But for action to occur, the signal must be passed onto other areas of the brain via the sine waves. The delta, theta, alpha and beta sine waves can spread to neighbouring ensembles of neurones.[2] The resulting minute electric fields in the CSF cause the voltage to fluctuate a little inside each neurone, which changes the firing of an AP to a particular place on the neurone's sine wave. This is called phase shifting and phase locking[3] (see Figure 1: Phase locking).

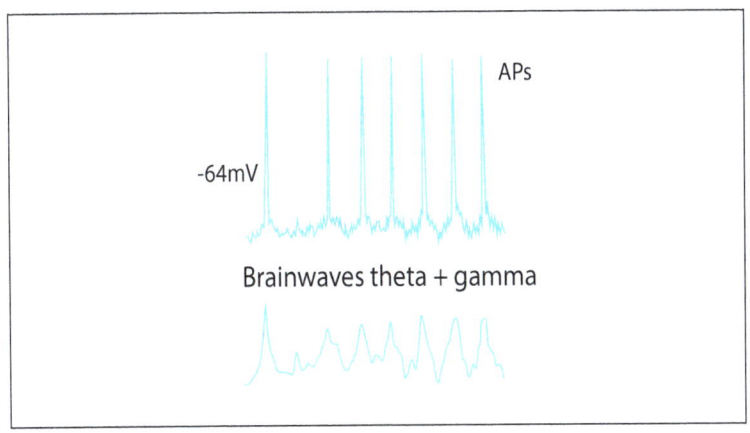

Figure 1a: Phase Locked Single APs

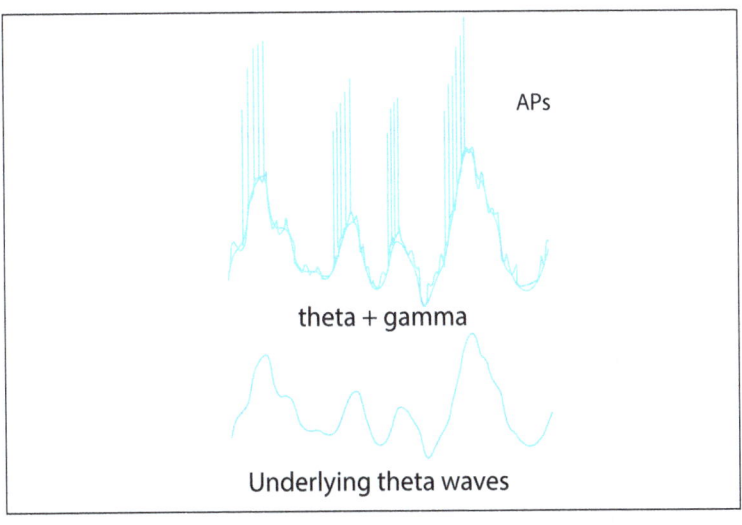

Figure 1b: Single Neurone firing bursts of APs

Figure 1a - 1b: Phase locking
1a: The APs lock onto the peaks of the underlying theta wave. 1b: The APs fire in bursts, locked onto the theta wave immediately before the peak.[3]

The phase locking spreads to some of the neighbouring neurones so that they all fire together at the same time and phase, therefore increasing the power of the signal. A code is produced by the firing patterns of the APs; the neurone ensembles go into overdrive and transmit the message to neuronal networks in different regions of my brain.

The sensory neurones in the parietal lobe signal the motor cortex, which commands my eye muscles to turn my eyes to the place on my leg where the March fly is biting. Now, the retina cells in my eyes send electric signals back to the visual cortex, where they are magnified and processed into an image. This in turn is transmitted to the temporal lobe, where it is recognised and named. The image and pain signals are collected in the brain's parahippocampus and forwarded to the brain's hippocampus, which is the centre for short-term memory creation.

So, now my brain knows **what** is on my leg – but I don't. This has all happened automatically at a subconscious level.

The March fly bite information journeys on to my brain's frontal cortex, where long-term memories are stored. The amygdala, which oversees the "flight or fight" response, also sends alarm signals to the frontal cortex. It then compares the inputs with its past memories of insect bites and collates all the information into a thought: "A March fly is biting my leg – swat that fly!"

Now, I am aware.

Multiple signals are sent to my motor cortex and most of the muscles in my body to position my head, neck, back, abdominal muscles, shoulder and arm, so that my hand can slap the fly on my leg. The movement of my hand through space is tracked by my parietal cortex and cerebellum, and my eyes lock onto the fly to see the result of my efforts. The whole episode is stamped with a time and place in the hippocampus and is relayed to my frontal cortex so that when my leg itches in the middle of the night, I will remember I was bitten by a March fly.

All these signals are activating many different areas of my cerebral cortex in a very short time span – much less than a second. How does this happen?

Microelectrode arrays on the brain surface of patients with epilepsy have shown theta waves occur all over the cerebral cortex[3] and there are five different theta waves in humans, each with a frequency ranging between 3–10 Hz.[4] Theta waves are thought to carry information between brain regions during thinking, learning and memorising. APs from different sensory inputs – such as vision, smell and hearing – are added onto one theta sine wave as it travels through the hippocampus.[5] We know from fMRI that different areas of the brain operate together at the same time and at the same frequencies. For example, theta power increased in the frontal and emotional cortex while people were doing mental arithmetic.[6]

So, now we know how the information is carried between areas in the brain. But what happens when it arrives at its destination?

The brain can "tune" itself just like a radio, so that neurones in some areas can receive the electrical signals from another area by "coherence".[7] This means that the frequency and phase of the sine waves of electric fields in two separate areas match perfectly[7] (see Figure 2: Coherence).

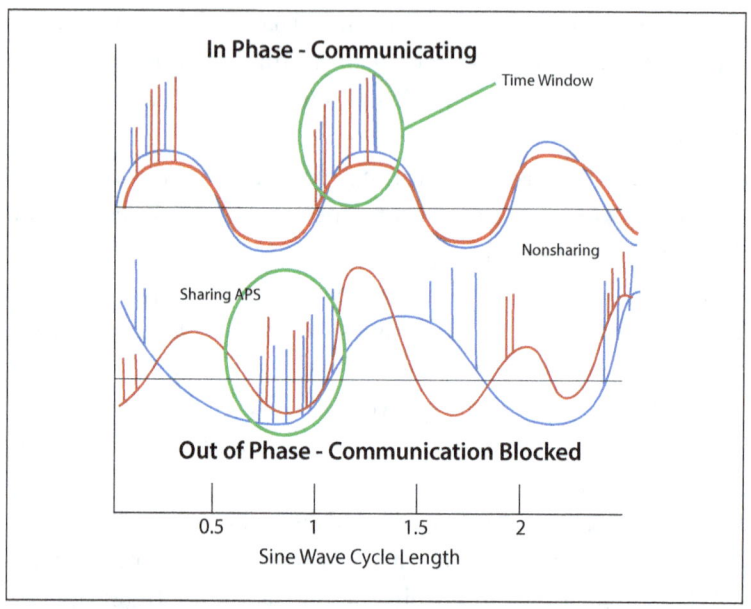

Figure 2: Coherence
The two theta waves (red and blue) can share APs with each other when they are in phase (upper diagram), but if they are out of phase by even half a cycle, they can't communicate at all (lower diagram).

The neurones amplify the signals through resonance. If a miniscule alternating electric current is injected into a neurone at the same frequency as its sine wave, this causes a large increase in the voltage inside the neurone, causing it to fire a burst of APs. This is like tuning your radio or TV to the correct frequency and then turning up the volume knob. Computer modelling has shown that when a weak stimulus from a few APs is combined with resonance, a signal can be forwarded to other brain networks.[8]

Although the original signal from the pain and touch receptors on my leg might be tiny, it can be magnified over and over as it passes from the "leg address" on my sensory cortex to all the other areas of the brain, and the billions of neurones that need to be activated so that I can swat the March fly.

Now we know how the brain receives and processes information from the outside world via our "normal senses" of touch, pain, sight and hearing. Let's move on to what it might mean if our brain could process information in another way.

11

Electrical Signals bypass the Normal Senses

Reflecting on how the brain receives signals led me to ask two questions. The first is: "Can the brain process information from the outside world directly, bypassing our normal senses?" The second, more important question is: "Can the brain receive and process information originating in another brain, without using our normal senses?"

Brain computer interfacing (BCI) has answered the first of these questions. The impetus for BCI research came from the dream of helping the blind to see, the paralysed to walk and amputees to use a robotic arm or leg. The initial research used computer software to analyse a person's EEG data as they moved a cursor on a computer screen. The machine learned the patterns so that the cursor moved when the person used thought alone. Now, the dream is realised – a person can think and activate a robotic hand to grasp and pick up an object.

The following research details how the brain can respond to input from an external electrical source. Alternating electric currents applied to the scalp changed the timing of APs.[1] An external alternating current at the same frequency as the brainwaves increased the power of the brainwaves, and APs were phase locked to the artificial electrical input and spread widely over the network by resonance.[2]

Going one step further - can external electrical stimulation carry meaningful information? Can experimental animals feel "virtual sensations" produced by electrical stimulation of the brain?

Let's look at studies of rabbits' whiskers which are very sensitive to touch, protrude so that they are equal to the width of their body and are used to measure the size of holes and burrows underground in the dark. Each rabbit has a "whisker address" in its sensory cortex and scientists applied alternating current to the rabbit's scalp overlying this address. But since rabbits can't tell us if their whiskers tickle, the scientists had to prove this was occurring in a round-about way.

Their left whiskers were tickled electronically and shortly after, air was puffed onto their left eye which made them blink. After training, the rabbits would blink after whisker tickling but **before** the puff of air was delivered. In psychology, this is known as a conditioned reflex. Then, the scientists applied alternating electric current to the "whisker address" of the right sensory cortex. The rabbits' response was the same – they blinked after the electrical stimulation to the "whisker address" but before the puff of air onto their eyeball. This proved that the rabbits had felt the electrical input as virtual whisker tickling – bypassing the normal sense of touch.[3]

Human studies have shown that brainwaves match and phase lock to electric current applied to the scalp, and this changes perceptions and behaviour. The signal from the motor cortex to a small muscle in the opposite thumb was almost doubled in strength when alternating current was applied at 20 Hz to the "thumb muscle address" of the motor cortex (20 Hz or a beta wave is the preferred frequency of the motor cortex).[4]

Long-term memories are formed during non-dreaming deep sleep, when slow delta waves (1–3 Hz) dominate in the frontal lobes. When alternating current of 1 Hz was applied to the foreheads of medical students just entering deep sleep, they were able to remember more paired words – such as chair/table – from a list they had learned the night before. Thus, external electric current was empowering thought![5]

Brain computer interfacing requires a computer to interpret brainwaves recorded by EEG.

EEGs of a person using a joystick to draw a pentagram on a computer screen were recorded and analysed by the computer, which was then able to reproduce the pentagram almost perfectly from the brainwave patterns alone.[6]

In a similar experiment, a computer analysed the brainwaves over the motor cortex as a person moved a joystick randomly in one of four different directions – left, right, up or down. Each direction had its own unique brainwave pattern. We "think" of a movement before it is performed, and the computer was able to predict the direction of the next movement from the brainwave pattern generated by the motor cortex.[7]

These experiments on animals and humans indicate that the brain can respond to direct electrical stimulation bypassing the normal senses. Furthermore, EEG recordings are not gobbledygook and brainwave patterns have meaning.

Brain-to-brain information transferral in animals

Which leads us to the second question: can one brain understand information that originates in another brain?

The following research is mind-boggling and demonstrates this is possible.

This experiment also involved whiskers but those of rats instead of rabbits. The sensations of one rat were sent virtually to the brain of a second rat, which then interpreted the information, decided between two possibilities and acted according to that decision.

The rats were trained to detect whether a "doorway" they had to pass through was narrow or wide by feeling with their whiskers. They then poked one of two buttons with their nose to show the size of the doorway. If they were correct, they received a small drop of water as the reward.

The rats then had microelectrodes implanted over the "whisker address" of the sensory cortex. A computer was used to magnify the brainwaves generated by the sensory cortex of the first rat, which represented the narrow or wide doorway. The second rat also had a microelectrode array positioned over its "whisker address". The computer transmitted the enhanced signal to the second rat so that it received a virtual sensation of the doorway.

The second rat then had to poke a button in its cage to show the size of the "virtual" doorway. It was correct 66% of the time – well beyond a result occurring by chance. The first rat got another drop of water if the second rat got the correct answer. If the second rat made a mistake, the first rat increased the frequency and strength of the brainwaves it was generating about its whisker sensations, so that it could receive the second drop.[8]

This experiment involved the transfer of sensations from the brain of one animal to another. Does this remind you of the twin who felt pain in her own hand when her sister's hand had been injured by the treadle sewing machine?

Brain-to-brain information transferral in humans

Experiments on humans have also been successful in transferring information from one brain to another without "normal" means of communication.

In one experiment, two people played a game on two computers in two separate buildings. A rocket flew across the screen which had to be shot down by a missile. The sender watched the rocket and imagined moving a cursor with his hand to the "fire" button. This involved "thinking" of the planned movement by the motor cortex. The first computer analysed his

EEG recording and transferred the brainwave patterns of the "fire" command to the second computer over the internet. The receiver did not view the game on his screen. When the "fire" signal arrived, the second computer stimulated the motor cortex of the receiver, causing his hand to drop onto a mouse pad which then fired the rocket. The transmission of these signals enabled the gamers to cooperate and shoot down the rocket. As the researchers concluded, a thought in the mind of one person was able to be transmitted to a second mind to bring about an action, albeit with the intervention of computer algorhythms.

In 1975, the Stanford Research Institute published results of experiments requisitioned by the CIA. The first of these used strobe lighting to flash a light at sixteen times per second for ten seconds into the eyes of a sender. This acted like a burst stimulus to magnify the intensity of the signal. The person acting as the receiver was seated in a room seven metres away that was shielded from light, sound and electrical signals The receiver was hooked up to an EEG which was tuned to the alpha rhythm of his occipital lobe. Any stimulation of any kind causes the alpha rhythm to decrease in power and become discombobulated. The light stimulus was alternated randomly with periods with no flashing light and the EEG measured the power of the alpha rhythm. When the sender was seeing the flashing light, the average power of the alpha rhythm on the receiver's EEG dropped by about 25%. The probability of this occurring was $p = 0.03$ so this was not a chance result.

A refinement of the experiment showed that the change in alpha rhythm only occurred in the right occipital lobe. This experiment demonstrated the transmission and reception of the light stimulus from the sender to the receiver, but the receiver was not aware of these changes in his EEG recordings – the signals received did not reach the level of consciousness.[10]

How do these experiments sit with telepathy?

These three experiments in an animal species and humans show that sensations can be transferred between brains by an invasive method in the rats but by a non-invasive and harmless method in the humans. But most importantly, the first two show that the patterns of brainwaves carry meaningful information and can be interpreted correctly by another brain.

12

A New Sense – the Magnetic Field

Our brain has a magnetic field and produces magnetic brainwaves. The sodium and potassium ions washing in and out of the neurones produce an alternating electric current which gives rise to an electric field, and therefore also gives rise to a magnetic field at right angles to the electric field.

The brainwaves produced by the magnetic field are identical to those of the electric field.[1] They are recorded using much more cumbersome equipment and the recording is called a magnetoencephalograph (MEG). The second joystick experiment described in Chapter 11 used MEG, rather than an EEG, to analyse the brainwave patterns and this proves that an MEG also is not gobbledygook.

Earth has a magnetic field which is generated by charged atoms of molten iron, swirling around in its liquid core, with the rotation of the planet. At the surface it is weak – two hundred times less powerful than the magnet pinning your photos on your fridge. The sun's contribution to our magnetic field is even weaker. The solar wind carries charged particles from the sun into our upper atmosphere, where they interact with atoms and molecules to form ions and the ionosphere. The movement of the ions in the ionosphere creates a small magnetic field which is strongest at noon and weakest during the hours of darkness, dawn and dusk.

It would be surprising if we were not influenced by – and did not use – Earth's magnetic field, as other living creatures do.

Salmon, green turtles and birds use the geomagnetic field to navigate through and over Earth's oceans to their breeding grounds.[2] And given the opportunity, dogs will align themselves north–south to pee and poo![3] The African mole-rat lives underground, is almost blind and orients its nests in the south–east corner of its huge underground dens by using the geomagnetic field.[4]

To date, scientists have not been very successful in identifying the sensory organs in animals that detect the magnetic field. Most of them found so far, contain magnetite, a special form of iron which can be magnetised and responds to small changes in the magnetic field, exists as crystals in certain bacteria, the skull of sock eye salmon and the beaks of some birds.[5] A magnetic receptor was not found in the mole-rats, but neurones sensitive to the magnetic field were discovered in a relay centre which also contained a GPS-like visual map of the outside world.[6]

Are humans sensitive to Earth's magnetic field? It seems so.

Studies have shown that geomagnetic storms bring on heart attacks, admissions to a psychiatric hospital for depression a fortnight later and, in South Africa, an increased suicide rate. Others found a link between epileptic seizures and the strength of the geomagnetic field.[7,8,9]

Early attempts to investigate the effect of the magnetic field using the electroencephalogram (EEG) found that changes in people's brain function depended on their physical orientation. Those sleeping in an east–west position, compared to a north–south one, entered dreaming sleep sooner. There were also significant differences in the EEG of people depending on whether they were sitting north–south or east–west when it was measured.[10,11]

In another experiment, humans were taken to a site deep in the forest, away from man-made electromagnetic radiation, and were exposed to artificial magnetic fields at delta wave (1.5 Hz) and alpha wave (10 Hz) frequencies. All the people tested experienced changes to their brainwave power in at least one of the lobes in their brain, but the responses were very individual. Two strengths of artificial fields were used, and the changes were greater with the stronger of the two and with alpha frequency. In one person, the alpha power decreased by half in the parietal lobes, was four times stronger in the frontal lobes and nine times stronger in the right occipital lobe in the weaker artificial field, compared to a staggering forty times stronger in the right frontal lobe in the stronger field.[12]

A similar experiment used artificial varying magnetic fields at theta frequency (7 Hz) but with significantly lower strength, to imitate the normal daily variation, or a stronger field to mimic geomagnetic storms. Theta power of the right parietal lobe was significantly increased with the lower strength but that of a magnetic storm suppressed theta power.[13]

In 1992, geologist Joe Kirschvink and his neurologist wife found magnetite in fresh post-mortem samples of human brains. The magnetic particles were extracted and found to be identical to the magnetite crystals found in fish and bacteria. Each gram of brain tissue contained 5 million crystals arranged in small clumps. The magnetite was distributed evenly through the cerebral cortex and the brain stem but the outer membranes covering the brain – the meninges – contained twenty times as much as the brain tissue itself.[14]

Magnetite was also found in fresh samples from the hippocampus of surgical patients.[15] More than twenty-five years passed before this finding was verified by the discovery of magnetite in human brains that had been preserved in formalin for over twenty years. The greatest concentration of magnetite was found in the brain stem of the preserved brains but unfortunately the meninges were not examined.[16]

We do not yet know whether the magnetite in our brain is located inside or outside the neurones. It is therefore impossible to understand how the magnetite interacts with magnetic fields, thereby affecting neurone function and brainwave power. We can safely assume though that it does!

In 2019, Joe Kirschvink's research team also proved that the magnetic field has a direct effect on the cerebral cortex. They placed their human subjects in a shielded room which blocked out light, sound and high frequency radio waves, so that the only sensory stimulus they received was from an artificial magnetic field of half the geomagnetic strength. They measured the alpha waves in the subjects' occipital cortex. Alpha waves are synchronised and strongest when a person meditates or sits quietly with eyes closed. If the person receives **any** sensory stimulus – such as touch, sound or vision – the alpha waves fade, lose synchronisation and become disorganised or totally discombobulated.

The research team moved the magnetic field horizontally or vertically. The alpha waves were discombobulated in some people when the field moved downwards or anticlockwise. The downward direction is the natural one for the magnetic field in the Northern Hemisphere, and the anticlockwise direction is the same as Earth's rotation.[17,18]

The few experiments quoted above clearly demonstrate that the human brain – like that of other animals – is sensitive to Earth's magnetic field and responds to changes in the direction and strength of the field with changes in the power of the brainwaves.

The brain and electromagnetic radiation

The neurones create both an electric and magnetic field and therefore must produce electromagnetic radiation, albeit of very small power.

The power of the electric field is diminished by interaction with the membranes, fat and cholesterol insulating the axons. However, the magnetic sine waves are not impeded by the brain tissue and can travel almost instantaneously anywhere within the skull. Therefore, it is probably the magnetic delta, theta and alpha sine waves which carry information from one part of the brain to another, bringing about the resonance and coherence of separated brain regions.[19]

The brain's magnetic field emanates from the inward folds of the cerebral cortex. These folds are also the location of the primary sensing neurones of touch, sight, and sound, where the neurones lie parallel to the surface of the cerebral cortex and Earth's magnetic field penetrates the brain without being altered or absorbed by the tissues of the scalp or skull.[19]

In the last four years, a new technology has been developed to measure the magnetic field within brain tissue. In an experiment using this technology, an array of micron-sized magnetic electrodes (magnetrodes) was inserted into the visual cortex of a cat and recorded the changes as the neurones fired, following stimulation of the cat's retina with flashes of blue light. Researchers found that the magnetic and electrical signals occurred almost simultaneously; the magnetic signal lagged by only 2 to 3 milliseconds. However, the power of the magnetic signal reached about 10 nanoteslas. This was huge – about 10,000 times stronger than the magnetic strength of the whole brain recorded by MEG.[20]

Differences in Earth's geomagnetic fields may explain differences in telepathic ability!

Why did the Cairns people seem to be more aware of telepathy than those in other countries, particularly Denmark and Norway? Maybe latitude was the reason. The geomagnetic field at the North Pole is twice as strong as that at the equator and we know that as the magnetic field strengthens, it can disrupt brain function. However, 33% of the people in Iceland and Great Britain had telepathic ability compared to 11% of the Danes and Norwegians.[21] These countries are all more or less at the same latitude, so that isn't the answer.

Earth's magnetic field varies from place to place, depending on the quantity of magnetic rocks and minerals in the planet's rocky outer layer.

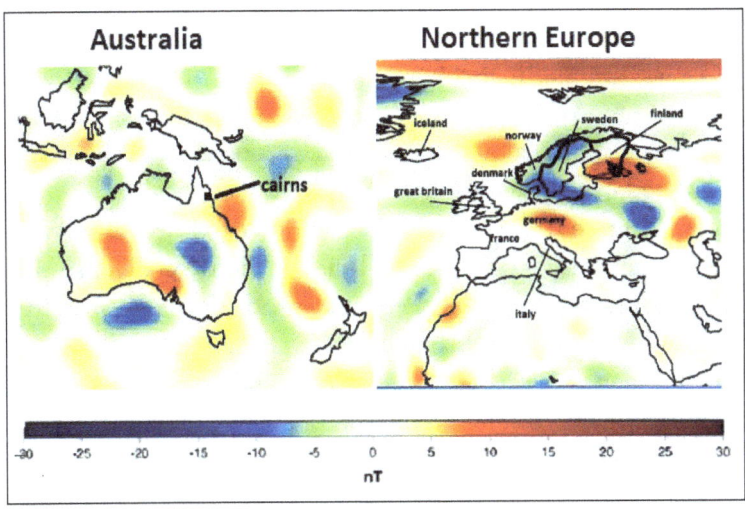

Figure 1: Geomagnetic survey of Earth's surface magnetic fields
The red and yellow areas show increases in the magnetic field in Earth's crust, with the field rising out of Earth's surface. In the blue areas, the field dives into the crust. (Adapted from Nasa Image).[22]

In Figure 1, you can see that in Norway and Denmark the magnetic field dives into the earth whereas Cairns sits next to a hot spot where the magnetic field flows outwards. Norway and Sweden have the world's largest deposits and mines of ferromagnetic minerals. These minerals change the magnetic field direction and strength, which in turn may explain the differences in telepathic ability.

A map of the world's surface magnetism shows that there are very few hot spots located in the inhabitable areas of the world (see Appendix/Section 4/Figure 3). The Louisiade Archipelago is close to that of Cairns; David Unapion's people lived in a hot spot in South Australia; Iceland and Britain border one in the North Atlantic and the Mediterranean countries are close to one in Western Europe, overlying Germany. These are the countries where surveys showed ESP and telepathy were more common. In the USA, where almost 60% of people surveyed by Andrew Greeley were telepathic,[21] there are three hot spots located along the Eastern Seaboard – one of which is centred over the Bermuda Triangle.

In summary, the human brain is sensitive to Earth's magnetic field which means that this is a newly discovered sense that we humans possess. The neurones produce brainwaves that are both electrical and magnetic in nature, and therefore create electromagnetic radiation – albeit of small power. In the next chapter, we will discover how the brain can send and receive information transported by electromagnetic radiation.

13

Is the Human Brain an Antenna?

For millennia, humans have used speech, the written word or body language to communicate with each other. This century, we have greatly extended our ability to communicate – at a distance and immediately – to Earth's inhabitants by using electromagnetic radiation. Both light and radio waves – carry information as electrical and digital signals to their receivers: mobile phones, radios, TV and computers.

Man-made electromagnetic radiation and signalling requires huge energy inputs and is very powerful. The frequency of a radio wave can range from 100,000 to millions of cycles per second. The magnetic field of an MRI scanner measures 1.5 teslas (T); Earth's geomagnetic field is measured in nanoteslas (nT), which are one billionth of a tesla; the magnetic field of the human brain is only one to two picoteslas (pT), or one trillionth of the power of an MRI scanner.

In 1975, the CIA was particularly interested to see if the human brain could influence the magnetic field. A second order magnetometer, usually used for measuring very small magnetic fields less than one picotesla – such as that produced by the human heart – was placed in a laboratory. It was connected to three measuring devices – an oscilloscope, a meter and a graph recorder. A remote viewing adept sat in another room four metres away and tried to "perturb" the readings on these instruments by increasing the strength of his magnetic field when given an "on" signal. He received feedback from the instruments while attempting this feat. The magnetic field increased in power during the active interference twice as often as when he was resting; the probability of this occurring was p = 0.004. This experiment was replicated elsewhere by another researcher, with the subject in a room fifteen metres away. This time, the signals were increased three times as often compared to the rest periods. The researchers were very careful to conclude only that this **probably** represented a perturbation of the magnetic field by human thought.[1]

We learned in the last chapter that Earth's geomagnetic field is generated by both the Earth itself and the sun. But Earth has another type of electromagnetic radiation, generated by lightning.

Two thousand thunderstorms occur on Earth at any one time, each producing fifty lightning strikes per second. This continuously bathes Earth in electromagnetic radiation. This radiation has a very low frequency – 1–300 Hz – and is called Extremely Low Frequency radiation, or ELF waves for short. ELF waves are special. They travel at the speed of light; have a very long wavelength; pass through our heads, buildings and mountains easily; and travel into the earth a short distance and for many kilometres through seawater. They have been used by the US navy to contact submarines. They are reflected by the ionosphere sixty or so kilometres above the ground, and so travel around the world in the space bounded by the ground surface and the ionosphere.[2]

As a bolt of lightning hits the ground, the electric field of the ELF wave spreads vertically. But the magnetic field spreads horizontally – in both the east–west and north–south directions – while the wave itself circumnavigates the world both clockwise and anticlockwise, several times in one second because of its long wavelength. Where the wave meets itself coming from the other direction, interference occurs to form standing waves where the power is amplified.

The fields of an ELF wave are very weak but have the same strength at all places around Earth simultaneously. The electric field measures up to 300 uV/m and the magnetic field is only 1 picotesla. The ELF waves have a range of frequencies but the most common and the most powerful is called the Schumann resonance. It has a frequency of 7.8 Hz, a magnetic power of 2 picoteslas and a wavelength of about 40,000 kilometres – the same as Earth's circumference. It loses its power very slowly, meaning it can circumnavigate Earth several times in one second, before it fades out.[1,3]

Does this remind you of something?

The strengths of the electric and magnetic field of an ELF wave and the human brain are identical, and the Schumann resonance matches the theta wave of the human brain almost perfectly. Here we have a "coincidence" of great improbability.

The human brain and ELF waves

The million-dollar question is: could ELF waves carry thought messages from one human brain to another?

The electric field of an ELF wave is only about 300 microvolts. This is not strong enough to penetrate the scalp and tissues to reach the neurones lying on the surface of the brain.

However, the magnetic field of ELF waves appears to pass unimpeded through the skull, as it does through earth and water.[4] It need not be any stronger than 1–2 picoteslas to deliver a signal to a neurone, and the anatomy of the brain aids the delivery of information carried by the magnetic field to the primary sensing areas of the cerebral cortex.

The sensing neurones of the tactile and visual cortex are in deep folds on the outer surface of the brain. They are aligned perpendicular to the surface, so they and their dendrites come to lie parallel to the surface of the brain, exposing their thousands of synapses to the horizontal magnetic field of the ELF wave (see Figure 1).

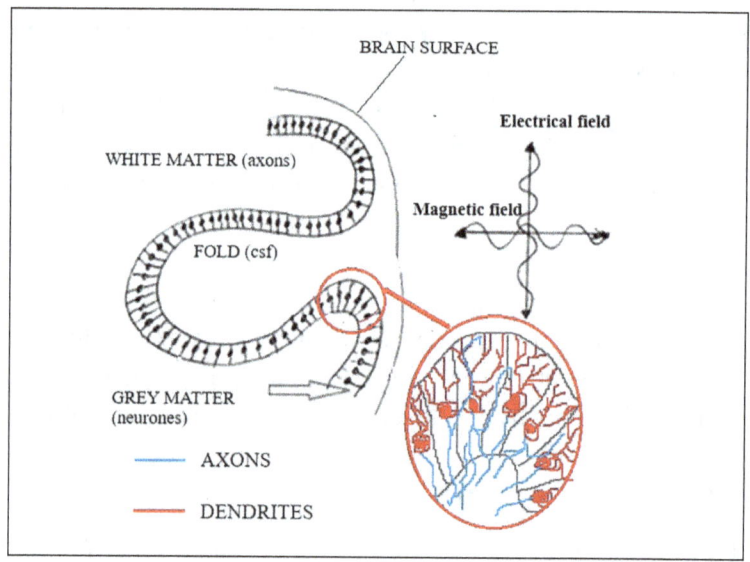

Figure 1: Alignment of neurones to surface of the brain

My physicist friend who suggested that thought messages could be carried by electromagnetic radiation also suggested that the brain was acting as an antenna. Antennas transmit radio waves as well as acting as receivers.

The curvature of the folds of the cerebral cortex and of the hippocampus, which is shaped like a seahorse, causes the electric and magnetic fields to be distorted. Instead of spreading as a circle, they are concentrated and beam out like the light of a torch, perhaps allowing them to escape a short distance to the outside world.[5]

Let's leave the brain for a moment here and see how an antenna works.

An alternating electric current is applied to an antenna, which then emits electromagnetic radiation at a specific radio frequency (called the carrier wave). The necessary components are a power supply; an oscillator to produce a sine wave of constant amplitude; an encoding mechanism – either varying amplitude (AM), or frequency (FM) or on/off digital signals called bits; an amplifier; and a tuner for the antenna. Orthogonal frequency-division multiplexing (OFDM) is a new method of digital signalling used in wi-fi networks, mobile phones and digital TVs. It uses carrier waves of multiple frequencies spaced closely together and over a range of about 100 Hz, so the bit stream is divided between the carrier waves. This allows information to be transmitted at a slower rate, making transmissions more accurate.[6]

Now, let's return to the brain.

Are all the components required for an antenna present in our brain? The answer is yes!

We have a power supply in the form of electrical energy derived from glucose, measured as 25 watts/second in the human brain. We have multiple oscillating local fields in the neurones and CSF, which generate sine waves at theta and other frequencies. Information is encoded onto the slow waves, such as theta, by the APs (action potentials). These are phase locked so that one cycle of a theta wave can carry multiple APs with varied information as a package. That just leaves the amplifier and the tuner to account for.

The brain's small loop antenna

The theta wave circuit travels through the hippocampus, parts of the primitive forebrain and the cerebral cortex located in the central part of the cerebral hemispheres.[7]

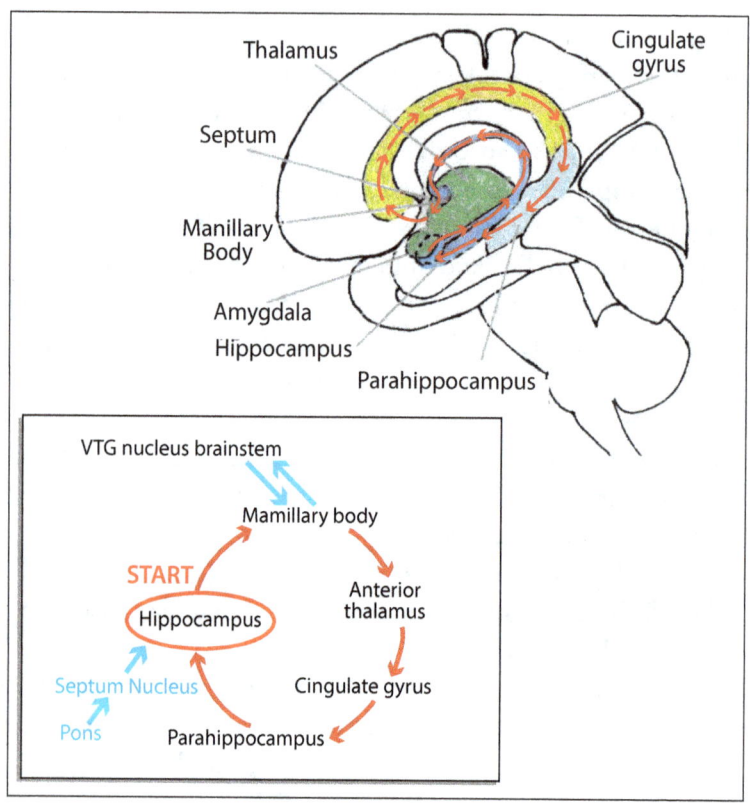

Figure 2: The antenna inside the brain

Theta waves commence in neurones in the pons – a part of the hindbrain. They are relayed through two other groups of cells in the forebrain, which modify the power and frequency and reach the hippocampus, then circle around through the anterior thalamus – part of the primitive forebrain. The theta waves then enter the cerebral cortex and travel` through the cingulate cortex – an association area for emotions and memory – to the parahippocampus and thence back to the hippocampus. The VTG nucleus is in the brainstem and regulates the theta rhythm to keep it in phase and time throughout the circuit – that is, it is acting as a tuner.[7] The amygdala acts as the amplifier, increasing the power of the theta waves as part of the fear response.[8]

The theta circuit is in the form of a small loop antenna (see Figure 2).

Loop antennas were developed by the US Navy to transmit ELF waves to submarines. These antennas transmit radiation by coupling with the magnetic component in the region of the antenna and then project the transmission in the plane of the loop. Loop antennas use ferrite (iron) to channel thousands of times more magnetic power through the antenna, to make it more effective. They are very sensitive to magnetic "noise" or static, however they can only transmit small amounts of data. The navy used them to signal to the submarine that communication was required.[9]

How does our loop antenna work?

The human brain also uses a narrow range of frequencies – like OFDM – over 1–100 Hz to transmit and encode APs into a "bitstream". This is transmitted throughout the cerebral cortex by the five different theta waves, plus or minus the other slow waves – delta, alpha and beta.

Information is carried to the hippocampus from the parahippocampus, the amygdala, the entorhinal cortex (involved with smell and the location of objects in space) and all parts of the cerebral cortex via a very large relay centre of the thalamus (the primitive forebrain). The brain's magnetic APs are almost 10,000 times stronger than the underlying magnetic field and can imprint theta waves as they travel through the hippocampus.[10]

So, both theta waves and coded information are funnelled through the hippocampus which stamps all our experiences with a time and place and forwards this to the frontal cerebral cortex – an area for working memory, long-term memories, planning and decision making.

The hippocampus contains magnetite, along with the brainstem[11] where the theta waves are initiated. This could increase the effectiveness of our loop antenna and lastly, a loop antenna transmits electromagnetic radiation in the "plane of the antenna", which means our thought waves would shoot out of our forehead between our eyes.[9]

Can ELF waves carry our thoughts?

In Italy, researchers collected ELF wave data and EEGs from patients over a three-year period. After conducting a comparison of the two data sets, they found coherence between the ELF waves and the theta waves of the EEGs in 17% of the recordings. (This may well have been a chance finding). However, further analysis of two patients' Quantitative EEG and an ELF wave simultaneously, using the same computer and software, showed that the ELF waves and brainwaves matched each other in power and phase at the theta, beta and gamma frequencies. The coherence was intermittent, occurring once or twice over a six-second period and lasting for about three hundred milliseconds – long enough for a thought to be transferred to the ELF wave in the hippocampus and carried out into the great wide world.[12]

One swallow does not a summer make, and this experiment needs to be repeated by other laboratories to see if this apparent matching of the ELF waves and brainwaves is real and repeatable. However, if it is true that the ELF waves and human brainwaves are matched in power and phase inside the brain, then it is possible for coded information in the form of APs to be imprinted on the ELF waves.

We have one last question to answer. How does the brain receive and interpret the "thought message" transmitted by electromagnetic radiation?

14

Interpreting Telepathic Information

If telepathic messages can be posted on ELF waves and carried to another, how is the message delivered and read by the addressee? The talent of "remote viewing" gives insight into how information can be delivered from one mind to another.

Remote viewing was first described by Upton Sinclair, a journalist and novelist, whose book *Mental Radio* was published in the 1930s in the USA. Sinclair would draw a small cartoon and place it in an opaque envelope. The following day his wife would relax, quieten her mind, try to "see" the drawing and copy it without opening the envelope. With practise, she became very accurate and was able to "see" the cartoon, even if the "sender" was forty miles away. Was she "seeing" or was she reading the mind of her husband or brother-in-law, who had drawn the original?[1]

The CIA became very interested in the potential use of psychic abilities for gathering intelligence. They were concerned that the Russians were researching ESP and telepathy – not only for spying but also to influence the behaviour of human populations. Consequently, they funded The Stanford Research Institute, initially to study telepathy and ESP. Hal Puthoff and Russell Targ were the Institute's lead scientists, and Ingo Swann – an American artist – and psychic Ray Price took part in "remote viewing" experiments. These were conducted with strict protocols to exclude errors and fraud, then were scrutinised by the CIA to determine their rigour and usefulness to the US military.[2]

In the experiments, two researchers would travel to a randomly selected location, at least thirty minutes away by car. Once there, they would concentrate on the scene before their eyes at a set time. Back in the laboratory, Swann or Price would then draw the scene and write comments about it. Swann received information about the scene as small fleeting impressions lasting a few seconds, over a fifteen-minute period.[3] (One of the viewings is shown in the Appendix/Section 4/Figure 3.) After repeating this process in a number of different locations, the drawings were given to judges, who independently visited each of the sites and attempted to match the drawings to each site. The results were good enough for the CIA to fund this research for another twenty years, which cost them many millions of dollars.

As time went by, some interesting facts were noted. Firstly, ordinary people were able to do remote viewing – including one of the administrative staff at the Institute. Secondly, remote viewers were able to draw the sites or some parts of them accurately but their verbal interpretation of what they drew was way off target. There was no apparent loss of information with greater distance between the target site and the remote viewer and lastly, isolating the viewer in a Faraday cage which blocks transmission of radio waves did not affect the ability of the remote viewers to gain information. All these experiments required the presence of an observer at the site – so the observer was thought to be sending visual and other sensory information to the remote viewers.

Later, sites which had no observer but were located purely by their latitude and longitude coordinates were used and successfully described by remote viewers.[4]

"Remote viewing" information is interpreted by our brain in the same way as the signals coming from our eyes.

Our normal sense of vision uses specialised neurones, called rods and cones, located in the retina at the back of the eye. These form a pixel-like image which is transmitted via the optic nerve and three relay centres to the primary visual cortex (V1), centred around a deep fold at the back of the occipital lobe (see Figure 2). The primary visual cortex is arranged into four network areas – V1 to V4 – and as the pixelated signal progresses through these, more details are added (see Figure 1).[5]

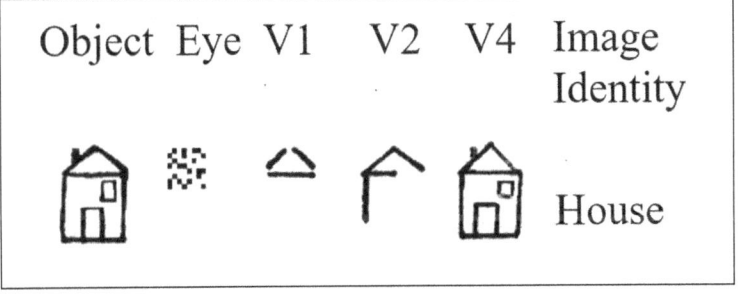

Figure 1: Brain processing of images
V1 identifies lines and edges. This pattern is then transmitted to V2, where more information such as curves and colours are added, before being sent on to V4, where shapes and complex details are added. Image credit:: Adapted from Cage, N. M. and Baars, B. J. 2018

From the visual cortex, the composite image heads off in two different directions.

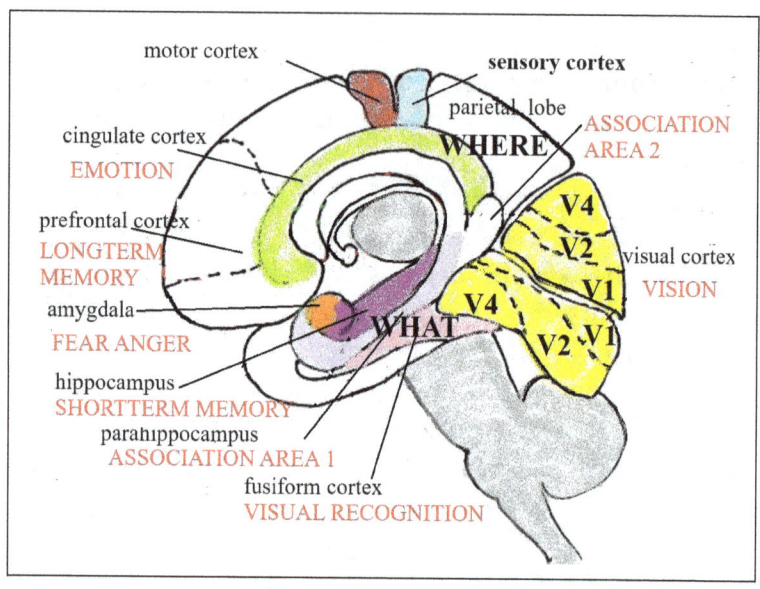

Figure 2: Visual pathways
The "where" stream travels upwards to the parietal lobe, which detects movement and the locality of an object in space. The "what" stream travels downwards to the temporal lobe where it is recognised as a house in Figure 1 above, and as a March fly biting my leg in Chapter 7. What happens next?

The parahippocampus adds other information to the image, before it is forwarded to the hippocampus and to Association Area 2, where autobiographical memory and navigation are integrated – such as remembering where you might have left your car keys. Finally, all this information is relayed to the frontal cortex which has two-way communication with the visual cortex, allowing information to travel back and forth between the two areas.[6]

Ingo Swann wrote a "how-to" book on remote viewing and the following diagrams are adapted from this book. These were the attempts from some early remote viewers in the 1890s and the early twentieth century, who were not using their eyes to view the images.[7]

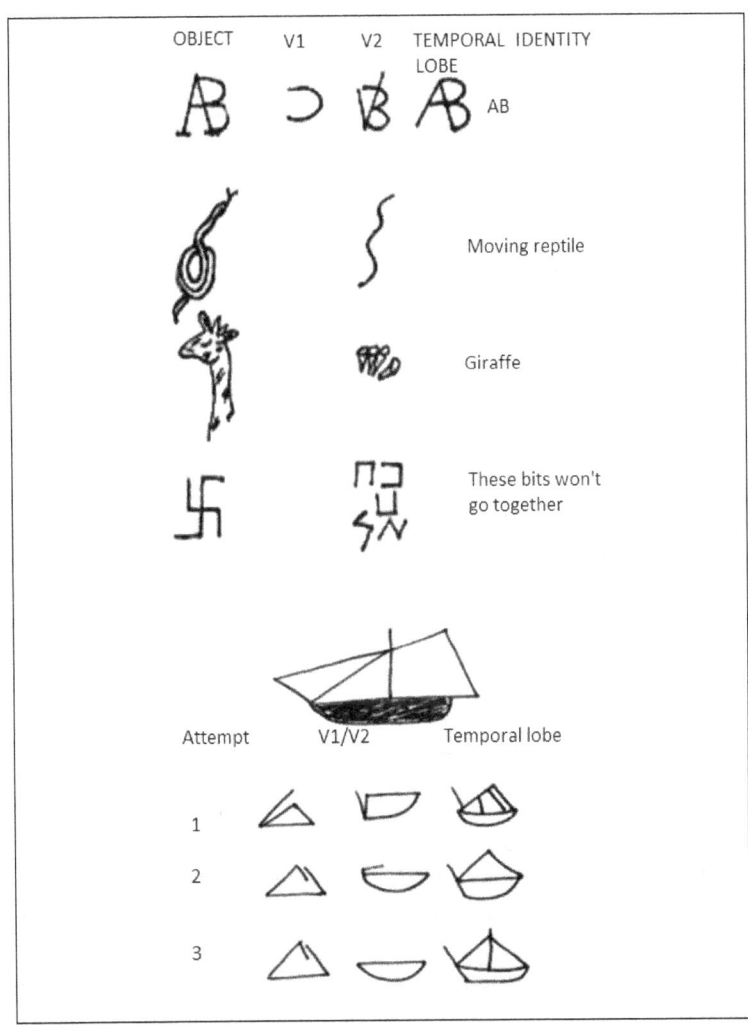

Figure 3: Telepathic viewing (by-passing eyes)
The remote viewing objects were received as separate parts which were assembled into a recognisable image and then identified – particularly the first remote viewing object AB and the boat. The parts of the swastika were processed but could not be identified. The snake was not only recognised but was shunted to the parietal lobe, where it was described as a moving reptile. Imasge credit: Adapted from Swann, I.1991

The only difference between remote viewing and normal sight is that the information is not entering via the retina but is arriving ready for processing at V1 – the primary visual processing centre of the occipital lobe.

Is this information carried by electromagnetic radiation directly to the neurones of the primary sensing area (V1) of the visual cortex, or is it arriving via the loop antenna to the frontal cortex and backwards to the visual cortex? It seems that direct stimulation of the neurones in V1 is more likely.

What is happening in the brain of remote viewers? Swann's EEG was recorded by the late Dr Michael Persinger – a professor of psychology at the Laurentian College in Canada, who had a life-long interest in the paranormal. Swann's EEG showed a large increase in theta waves in both occipital (visual) lobes – the right more than the left – while remote viewing. The accuracy of his drawings correlated positively with the activity of the theta waves – the more theta waves, the more accurate the drawings.[8]

In another experiment, an Indian "mentalist" was scanned with MRI while he was performing telepathy – again involving the accurate sending and reception of a simple drawing. The right parahippocampal cortex was active during the telepathic episodes.[9]

In the early twenty-first century, a famous Trinidadian psychic called Sean Harribance gave readings about photographs his subjects held of their relatives, while sitting close by in the same room. Though he had never seen the photographs before and did not know the relatives, Harribance apparently perceived quick images about them in his left visual field, which were processed by the right side of his brain. An EEG – and later a qEEG

– recorded his brainwaves continuously, to measure the power of the brainwaves and their location.

The recordings showed a specific pattern when he was "mind reading", lasting for about twenty seconds, and there were about six recurrences of this pattern per reading. The number of these episodes and their duration correlated with the accuracy of his reading, like Ingo Swann[10].

There was an increase in delta (1–3 Hz) and theta power (4–8 Hz) over the entire cerebral cortex, but it was especially strong in the right temporal lobe when he was "reading", compared to resting. Also, beta waves at 20 Hz were prominent and localised to the right parahippocampus.[10] This was the same area activated in the brain of the Indian mentalist.

There are other interesting findings from these studies.

Firstly, the researchers made EEG recordings of the subjects during the mind reading. The patterns of their brainwaves were almost identical to those of Harribance, and the qEEG results found that power of the beta and gamma frequencies of their left temporal lobe increased and matched that of Harribance's right temporal lobe. The matching lasted for about 130 milliseconds – the duration of one theta cycle or maybe a thought (see Figure 4/ Appendix/Section 4).

Secondly, the researchers measured the geomagnetic field alongside Harribance's head, near to the temporal-parietal lobes. The field on the right side of his head was 2000 nT – smaller than that of the left side. During the mind reading, there was a further drop in the magnetic field of about 150 nT after a fifteen second delay, which returned to its previous value when he stopped reading and when his EEG lost the special pattern.

Photons were simultaneously found in the space near his head, while he was mind reading.[11] It seems that the human body emits photons throughout the day because of changes in our energy metabolism.[12] It is possible that Harribance was drawing on the Earth's geomagnetic field, and/or receiving electromagnetic radiation carrying information from his subjects resulting in the emission of photons into the space adjacent to his right temporal lobe.

We can now understand how the brain can interpret a signal or information it receives from another mind. The signal is processed directly by the brain in the same manner as that coming from a sense organ, but without the involvement of the sense organ.

Using the limited data collected from a very small number of gifted individuals, we can conclude that:

- The right side of the brain – especially the temporal lobes and specifically the right parahippocampus – seems to be associated with ESP ability.
- Theta waves are important in this process.
- A specific pattern of EEG activity was recorded over the right temporal and frontal electrodes during telepathic activity, and this correlated with accurate information being disclosed about the subjects.
- Electrical activity measured by EEG in the brain of the mind reader and those whose minds were being read matched in time and power.
- It seems that the energy or information necessary to drive this process may be obtained from Earth's geomagnetic field.
- Images or visual signals are instrumental in conveying thought messages.

This doesn't explain how I received the messages about my friend's babies or my colleague's death. It is unlikely these came as visual images or that my other friend received the words "optometrist" and "appointment" this way. Maybe, the primary processing area of the auditory (hearing) cortex located in the temporal lobe was stimulated directly by the thought wave, thus bypassing my ears. Alternatively, my brain antenna received a coded message via an ELF wave.

In an experiment, Kevin Saroka – behavioural scientist and colleague of Dr Persinger – transferred the magnetic patterns of emotional words directly into the auditory cortex. The recipients were asked to classify the words into one of four groups. They found the words describing negative emotions such as "anger" were accurately selected almost twice as often as other words, and this was not occurring by chance.[13]

Alternatively, the antenna circuit of my brain received the information about my friend and colleague directly from an ELF wave. The information was then processed and relayed to the frontal cortex and entered my consciousness as an inserted thought or a "knowing".

There are two other characteristics of a telepathic thought to be investigated: recognition of the person sending the thought and the role that strong emotions play.

We use our normal senses to identify people, each of whom is unique – even identical twins. Individuals have hidden identifiers: their DNA, iris patterns and fingerprints. Astonishingly, we also have a "brain fingerprint"!

In 2015, scientists were investigating the connectivity of areas of the brain while their subjects were involved in a word memory task. They used qEEGs tuned to delta, theta, alpha, gamma and beta frequencies to construct maps of brain activity. The subjects used similar parts of their brain but each of the twenty-one subjects had a unique pattern. When they re-examined fourteen subjects six months later, the computer was able to identify each individual solely from their qEEG patterns. It's no wonder then that we can identify the person sending a telepathic message, or are more likely to communicate with someone we know well.[14]

Emotions play an important part in how our brain responds to the situations we experience in life. The amygdala is part of the emotional brain circuitry and is responsible for the "fight or flight" response when we encounter something frightening. You can see from Figure 2 that the amygdala is next door to the hippocampus and close to the parahippocampus. When we are alarmed, angry or afraid, the amygdala increases the power of the theta waves it produces, which in turn increase the power of theta waves in the parahippocampal gyrus. If we postulate that the theta waves carry our thoughts, then we can see that anger or fear amplifies the power of the thought and may allow it to be broadcast telepathically.[15]

15

The Hypothesis in a Nutshell

This hypothesis proposes that information is carried from one mind to another by electromagnetic radiation and that the human brain can transmit, receive and make sense of such information.

Prior to the 1930s, electromagnetic radiation was considered the likely mechanism that transmitted telepathic messages from one mind to another. However, Russian scientists discovered that a shielded room like a Faraday cage – constructed of copper and steel which blocks most electromagnetic radiation, such as ordinary radio waves – did not prevent the transmission of telepathic messages. This appeared to rule out electromagnetic radiation as the carrier of telepathic messages. However, the researchers subsequently found that very low frequency electromagnetic radiation (an ELF wave) is **not** blocked by such shielding.[1] The ELF wave can easily pass through buildings and other seemingly solid structures, and even partly into the earth with very little loss of its power. The attenuation of an ELF wave's magnetic field at 60 Hz is only 3 decibels, while that of an electromagnetic wave at 15 kHz is 68 decibels, so the ELF wave can travel around the Earth a few times without losing its power.[2]

Russian scientist I.M. Kogan proposed that very low frequency electromagnetic radiation could be generated by the brain and used to transmit thoughts, or patterns of information with a low bit rate, around the world. Further to this, if these patterns were carried via external, naturally occurring electromagnetic radiation, transmission would require less power than if the brain had to produce its own carrier wave.

Dr Michael Persinger took the hypothesis one step further by suggesting that it was the ELF waves at the Schumann resonance of 7.8 Hz and its harmonics which were capable of transmitting information from one brain to another.[3]

This hypothesis has been neglected for many years, mostly because there was so little information about how the brain processes chemical and electrical impulses into thoughts. Scepticism about ESP, including telepathy, has been rooted in the fact that there has not been any acceptable theory to explain its mechanism of action, compatible with current physical and biological principles.

In the last twenty years, there has been an exponential explosion of neurological research. This has delivered new insights into how neurones and the brain function, therefore enabling the resurrection and formulation of Persinger's hypothesis.

The following properties of a telepathic message fit with the idea that the brain can send or receive telepathic messages using electromagnetic radiation:
- The telepathic message travels very fast and can be almost instantaneous.
- The telepathic message can travel a very long distance between two minds.
- The detail is more accurate if the two minds are in close proximity.
- Conversely, the message is vague with greater distance between the minds.

- The sender of the message can be identified by their "brain fingerprint".
- The sender's message is amplified if they are emotionally distressed.
- The receiver of the message needs to have a quiet mind.
- The receiver of the message recognises the thought as foreign.
- Telepathic messages may be transmitted directly to the primary processing centres of the cerebral cortex, or via the theta wave antenna, to arrive as a foreign thought or "knowing" in the frontal lobes.

The hypothesis is supported by the following:
- Brainwaves are meaningful; they can be interpreted by brain computer interfacing and can transfer information from one mind to another. The information carried by the electrical and magnetic brainwaves is identical.
- A theta wave circuit passes through the parahippocampus and hippocampus, where coded information from all senses and areas of the cerebral cortex is added in the form of APs. It is stamped with a time and place and forwarded to the frontal lobe for storage as long-term memory, origination of thought and decision making.
- The theta wave circuit forms a loop antenna, sensitive to the magnetic brainwaves, which can be tuned and amplified and project low frequency electromagnetic radiation to the outside world.
- ELF waves have the same power (up to 300 microvolts and 1 picotesla) as the human brain.
- ELF waves pass through the human brain and cohere intermittently with human brainwaves for up to 300 milliseconds – long enough for a thought to be

transferred to the ELF wave and carried out into the world.
- Images received during remote viewing are processed by the visual cortex, in the same way as images received by the retina of the eyes.
- EEG and fMRI show that the right temporal lobe and parahippocampus seem to be the areas of the brain where telepathic messages are received.
- The magnetic field varies in its power and may be used as a source of energy for the reception of telepathic information.

From an evolutionary point of view, telepathy is an extremely useful adjunct to our normal senses, in that it has survival benefits for those who possess this ability. It would be very surprising if living organisms and humans did not use this talent.

Currently, this hypothesis is an idea with compelling but circumstantial evidence. It needs to be investigated and researched by scientists with curiosity and an open mind, to either conclusively disprove, or to validate it.

Part III: Mysterious Extrasensory Perception

This hypothesis of telepathy can explain some of the experiences of my patients, but those that involve knowledge of events requiring a distortion of time remain a mystery. This section of the book commences with visitations from loved ones at the time of death – which **can** be explained by telepathy. The book will then progress to examples of the most common experiences, such as clairvoyance and seeing or feeling ghosts. Rarer events, such as out-of-body experiences and psychokinesis, are included because they have also been investigated and correlated with detectable changes in brain activity.

16

Visitations from the Dying

Awareness of the death of a loved one was very common in my patient sample. Eighteen people disclosed these happenings, under varying circumstances. A few were awake at the time, some were woken from sleep, some dreamt, while others experienced bizarre communication through man-made appliances, such as the telephone.

Occurrences while wide awake

- Carol's four-year-old nephew turned to his mother one day and said, "Grandma just died." She had indeed died – elsewhere.
- Natasha's aunt was "psychic" and was dying of cancer. She promised that she would come and say goodbye to Natasha when her time came. Natasha was cleaning her house when she heard her aunt's voice say, "Bye lovie" at 2.30pm – the exact time of her death.

Occurrences that awoke patients from sleep

- Pat's aunt, whose son was serving on a destroyer in the Navy during the Second World War, woke up and said, "Bob has just been killed!" His vessel had been blown up and sunk at that moment.
- Anne's story is a little different. Her father was in hospital with a flare up of rheumatoid arthritis. He was receiving

treatment but was not considered to be seriously ill. Anne woke in the middle of the night and felt her father sitting on her bed. She knew it was him. She felt overwhelming sadness and thought, "Dad has died." The next day, she found out this was true.
- Wendy had recently become engaged. Her fiancé suffered a heart attack at her home early one evening and was taken to hospital. She woke in the middle of the night and saw him standing at the foot of the bed looking at her. At the same time, her usually quiet and well-behaved dog was inside, barking its head off. Her fiancé didn't speak and gradually faded away. The hospital called Wendy early the next morning to tell her that he had died overnight.

Dreams of loved ones

- Alice, her grandmother, mother and aunt all had a very vivid dream of Alice's great-grandmother on the night she died. On waking next morning, they each found their feet were sticking out or through the bedrails at the foot of the bed.
- Mandy was very close to her grandfather and she dreamt of him two days after he had been sent home from hospital. When she woke in the morning, she couldn't shake off the feeling that something was wrong. She and her sister were at home when the phone rang. She immediately started to scream and cry before her sister took the call, because she knew it was someone calling to tell them of their grandfather's death.
- Yi was pregnant when she had a strange dream about her paternal grandmother, who lived in China where Yi had spent her childhood. She dreamt her grandmother had phoned and invited her to a family dinner. When

she arrived, only her grandmother was present; the table was set but the plates were all empty and there was no food. While dreaming, Yi recognised this as being very strange and the memory of this dream worried her for a long time. Many months later, after she gave birth to her daughter, her parents told her that her grandmother had died about the time of her dream. She was extremely angry that her parents had kept this secret, but they were very superstitious and worried that the news may have a bad effect on her pregnancy.

Warnings of imminent death in dreams

Two stories are related to warnings that the loved one was going to die.

- Helen's husband dreamt that he needed to make a cross for his mother from light and dark wood, and take it to her. The following day they discovered that she'd had a heart attack and had died overnight. Even though it was too late, he made the cross for her and placed it in her hands at the funeral parlour.
- Elizabeth and her husband were estranged from his mother but on a Sunday in April, Elizabeth told her husband to visit his mother because it may be his last chance. A few days later, she dreamt she saw someone in the distance and said, "Who's there?" The person came closer – a little too close for comfort – and she recognised her mother-in-law. She was using a walking-stick; her hair was sunlit and shining and she looked very happy and free of her usual anger. Elizabeth felt the unspoken message was: "I am going." Her mother-in-law died the following Sunday – sadly without reuniting with her son.

Odd electrical events associated with death of a loved one

- Debbie's sister died of cancer and Debbie woke in the middle of the night to find the Christmas lights flashing.
- Judith's niece was living in her grandmother's house. Not long after her grandmother died, the telephone rang and the niece heard her grandmother saying, "Don't worry – I'm in a good place!"
- Lynne's younger sister, whose husband died of melanoma, used to sense him nearby and smell him. One night his boat, which was housed in the garage, started up of its own accord.

I have saved the most amazing experience for last and this has a bit of everything!

Eva was born in the Caribbean and her family migrated to the UK when she was young. Eva has had many psychic experiences over her lifetime but this one concerns the death of her mother.

It was Eva's nineteenth birthday and she was on the bus to the city to meet some friends for a coffee before work. The bus passed a street which was blocked off with yellow tape, police cars and ambulances, and Eva "knew" that her mother had been killed in a traffic accident in this street. She wanted to get off her bus, but something prevented her. So, she travelled into town and told her waiting friends that she had to phone Accident and Emergency. She was advised to go to the hospital, where she found out that her mother was the person in the accident and that she had been killed.

At the time, Eva's younger brother was boarding at a school some distance from London. Her father phoned the headmaster and advised him of the death, but requested that his son not be told so that the family could visit and break the news to him in person.

Eva, her father and siblings arrived later in the day and were escorted to the headmaster's office, where they were served afternoon tea. Her brother was called to the office. As Eva tells the story, a very untalkative, surly teenager entered the room and the family gently told him the news of his mother's death.

"Yes – I know!" Eva's brother yelled. The family and headmaster were very upset that the news had leaked somehow and asked him how he knew.

"Well, I was in the shower this morning and the phone rang and one of the other boys answered and called me because Mum was on the line. So, I had to get out of the shower dripping wet and take the call."

She said, "I've been in an accident but I'm okay and I just want you to get on with your life."

The family were amazed because Eva's mother had been killed instantly. This was well before the days of mobile phones, and it was impossible for her to have phoned her son. "Was it really her?" they asked.

"Of course, it was," her brother replied. "I know the sound of my own Mam's voice – don't I?"

Eva's mother communicated with her telepathically, at or after the time of her death, and with her son electronically.

Most of these experiences can be explained by telepathic messaging between the dying person and the recipient. However, for the dying person to be the sender, the neurones of their cerebral cortex need to be active to generate a thought wave. Impossible as this may seem, there is recent evidence to suggest that this may be exactly what happens.

In 2013, a landmark experiment examined the activity of the brain in rats immediately following an induced heart attack and cardiac arrest. EEG electrodes monitored the frontal, parietal and occipital lobes, and the readouts were analysed. The rats' brains went into hyperdrive with a surge of synchronised gamma waves of low-to-high frequency, lasting for up to 30 seconds. These gamma waves were strongly coupled with theta and alpha wave phases, and there was a huge increase in the interaction between the frontal and occipital lobes. This activity far exceeded that of the normal waking state and was evidence of heightened information processing within the brains of all the rats at the time of death.

The studies showed that feedback of information from the frontal lobes to the occipital lobes – that is, memory to visual cortex – was eight times greater than during the waking state. Similarly, trafficking of information from the visual cortex to the frontal cortex was five times greater. In humans, signals from the visual cortex (the occipital cortex) to the frontal cortex are subconscious but signals in the opposite direction are associated with conscious perception – that is "seeing".[1] This evidence of highly-organised brain activity consistent with conscious thought could explain the highly lucid and realer-than-real mental experiences reported by near-death human survivors. Many near-death survivors report a review of their lives which appears as a series of images, as if they were watching a movie.[2]

There is some limited evidence that human brains are similarly active at the time of – or shortly before – death. Seven patients, who had normal brain function but were dying, were monitored while their life support was withdrawn. Their brain function was monitored using an EEG electrode, but instead of the normal read out, a general index of consciousness was computed on a scale of zero to 100. These machines are commonly used in anaesthetised patients, to monitor their level of awareness.

After the patient became pulseless and blood pressure dropped, the signals on the machines declined to a lower level. But this was followed by a spike on the readout approaching

80 to 100 – levels associated with conscious thought. The spikes lasted from thirty to 180 seconds, depending on the patient.[3] What we can surmise is that the brain of a dying person is sufficiently active to broadcast a thought message just before death.

For the first time, an EEG recorded the brain activity of a patient with a head injury as he transitioned to death. There was a burst of gamma waves coupled with alpha waves, most notably in the left cerebral hemisphere, just before he died. In healthy people, these are the frequencies involved in thinking and memory recall.[4] This finding supports the idea of end-of-life recall and the ability to transmit messages telepathically at the moment of death.

There is another possible explanation for being aware of a loved one's death. What if those who are linked telepathically, subconsciously monitor the presence of each other throughout their lives, and the absence of the unique brain pattern of the dying person alerts one of the pair to that person's death?

To illustrate this, I will share one of my own experiences that puzzled me for years, until I started researching information for this book.

While I was completing my studies for my science degree, I lived with my family in Holland Park, Brisbane. One night, I felt a strong need to talk to my boyfriend. He lived in college at the University of Queensland, in St Lucia. I phoned the college but was told he wasn't there. As the night got later and later, the need to speak to him intensified, although there wasn't anything particular to discuss that couldn't wait until the following day. Eventually, I felt that I absolutely had to find him.

I set off at 11pm in my unreliable old car, to travel the twenty kilometres to the university. I first went to his college, to look for his car. It wasn't there, so my next port of call was the Chemistry Building. I found his car parked outside. When I ventured in the whole building was empty and in darkness except for a thin line of light showing under the closed door of his lab. His laboratory had been an old store-room, so it was very small with just room for a bench with a sink. The ceiling was high, but the room had no ventilation or windows. My boyfriend was completing his Honours degree in Physics and was studying gas chromatography. He dissolved chemicals in ether and then put the samples into the gas chromatography machine and read off the electrical signals and patterns produced.

When I opened the door to his lab, an overpowering smell of ether hit me. I found him unconscious, slumped over the bench, his nose very close to a large pool of ether spilled across the black wooden surface. More ether was in a large beaker without a lid close by. I soaked up the puddle and poured the rest of the ether down the sink. Gradually, my boyfriend regained consciousness but was very groggy. I got him back to his college and he slept it off; neither of us realised at the time how close he had come to dying. I got home well after midnight to face my mother, who was enraged by my driving off in the middle of the night to visit my boyfriend.

Ether is an anaesthetic and has serious side effects in overdose. It has been replaced by more modern inhaled anaesthetics and is now rarely used in Western countries.

Research into the effects of various anaesthetics on brain function has shown that ether and similar gases cause the theta and gamma waves to disappear from the EEG tracing. These are the frequencies most likely to produce a telepathic thought wave.[5] Once I knew this, it became clear that as my boyfriend had drifted into unconsciousness; his "brainwave fingerprint" had switched off. I had no understanding of this at the time, but I was sufficiently alarmed to go searching for him when his "signature" was suddenly absent.

I wonder if this is how Sibyll "knew" that her son had had an accident, when she started praying for God to take her instead of him. Was he unconscious and "off the air" and this alerted her to his plight?

Did Mandy "know" that something was wrong the morning after her grandfather died because she couldn't sense his "brainwave fingerprint"?

It seems to me that it is important to find out what happens within the dying brain. If enough volunteers are willing to be monitored with full EEG studies at the time of their death, we could clarify this process.

17

Brainwaves and ESP Phenomena

Two of my patients had **out-of-body experiences (OBEs)** as young children.
- Suzie said that as a child of five or six, she could fly up out of her body whenever she desired during the daytime. She could see her body down below and was able to travel around the neighbourhood and watch her neighbours from above, as they went about their day-to-day activities. She used to call out to her father if he came outside: "Look at me – I'm up here!" As she became older, she says she grew out of it.
- Kathryn told of the same thing in almost the same words. She said that until she was eight or nine years of age, she was able to fly up above herself and look down and see herself standing below.

A neurologist called Wilder Penfield conducted many experiments on his patients with epilepsy. He would probe the patients' brains with electrodes in the operating theatre while the patient was awake, to identify the location of the damaged part of the brain causing the epilepsy. The probes allowed him to map the cerebral cortex and find what things happened where. One of his patients felt as if she was floating out of her body, when he probed the area at the junction of the parietal and temporal lobes, that is, she experienced an OBE.

In 2002, another epilepsy patient experienced an OBE with stimulation of the right angular gyrus – also at the junction of the temporal and parietal lobes where sight, sound and feelings are processed to be relayed to the frontal cortex.[1]

Recently, psychologists at the University of Ottawa were able to perform a functional MRI scan on a healthy adult female student, while she voluntarily left her body to float or move above it. They found that two areas of her left temporal lobe – the temporal parietal junction, along with part of the motor area of her brain – and the cerebellum were active on the scan while she was "out of her body". These were different to the areas that lit up when she was asked to simply imagine doing this, or when she was asked to perform movements of her fingers or imagine playing a game of jumping jacks.[2]

So once again, we find that the temporal lobe and the parietal temporal junction seem to be involved with a very different kind of extrasensory perception.

Telekinesis – or psychokinesis – is the movement of objects without physical interaction.

My siblings and I had one of these experiences when we were teenagers.

We were playing with a homemade Ouija board on a rainy, grey Sunday afternoon. The Cold War was in full swing, with Russia and the USA facing off over the Bay of Pigs in the Cuban Missile Crisis and "ban the bomb" was graffitied onto the railway underpass into the city. We were worried that the end of the world was coming, and my brother would be called up to war. As we giggled and bickered over who was responsible for pushing our father's shot glass around the board, a small vase suddenly "jumped" off the bookshelf and fell to the floor. We were scared stiff, covered in goose bumps, and rushed out of the room to throw our Ouija board into the rubbish bin.

Four of my patients had stories of this type.
- Layla's niece, Cecilia, lived in an old house when she was in her early twenties. Each time she went out the kitchen

door, she would experience a strange sensation, and cups and saucers would launch themselves off the kitchen shelves to fly through the air.
- Jeannie was devoted to her elderly mother and was very upset when she died in her nineties. For a time after her mother's death, Jeannie would sense her presence, and a small carved wooden cat would fly off the top of the fridge.
- Roberta had a pair of misbehaving doors in a cupboard in her pantry. No matter how many times the doors were closed with various fasteners, they were invariably found open again shortly after. She never observed the doors opening.

All these stories have an underlying similarity. The objects do not topple over and roll or fall off. They leap, jump or fly before crashing to the ground, as if a strong force has acted upon them to make them airborne.

If I hadn't seen the small vase leap off my bookshelves to crash on the floor as a teenager, I would find these stories very difficult to believe.

A mentalist called Guy Bavli can be seen performing a feat of telekinesis on "Stan Lee's Superhumans" on YouTube.[3] He "tipped" a pen which was resting on the top of a wineglass into the glass without touching it. Psychiatrist Dr Thomas Brod and his colleague Bill Scott monitored the qEEG of the mentalist during this demonstration, by attaching electrodes to either side of his forehead. They kindly provided me with the academic poster detailing their results.[4]

Usually, an EEG screens out the frequency at 60 Hz, because this is the frequency used by electrical appliances in our buildings. They deliberately did not screen out this frequency, since any external gadget used fraudulently by the mentalist to perform this feat would show up in the EEG.

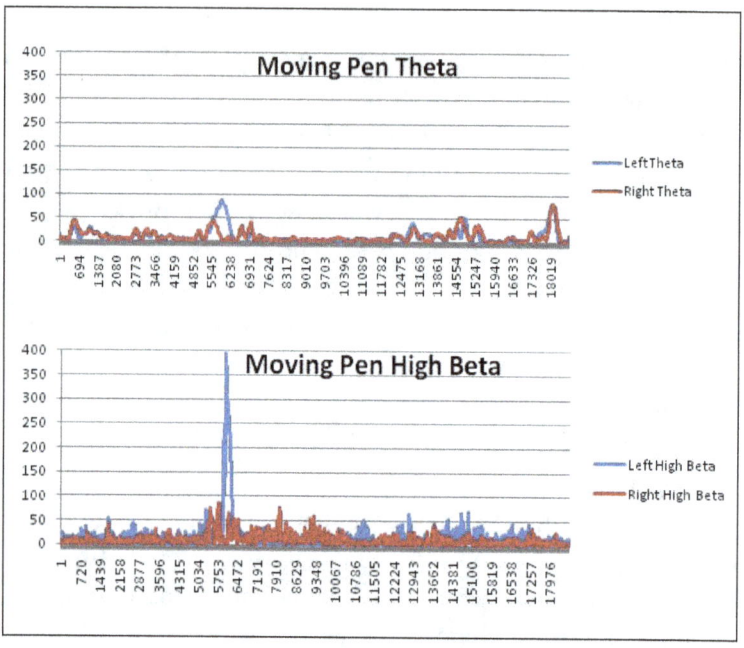

Figure 6: EEG recording frontal electrodes during telekinesis
Image credit: Brod and Scott 2011 [4]

In Figure 6, the blue trace is from the left frontal electrode; the red trace is from the right frontal lobe. There was an obvious increase in the power of the theta, beta and gamma waves in the left frontal cortex at the time when the pen tipped into the glass. The theta and gamma waves were almost synchronous, but the beta wave lagged by 200 milliseconds or so. The increase in power was almost 200 times stronger for the gamma waves compared to the mentalist's EEG at rest and even more for the beta waves. These values were twice what would show in a patient having an epileptic seizure. In other words, these brainwaves were very powerful.

There was no sign of any activity at 60 Hz, indicating there was no electrical interference from outside. The mentalist did not touch the pen physically. The video on YouTube shows that he waved his arms around a lot before the pen fell into the glass. Large muscle movements produce quite large changes in the magnetic field surrounding a person, which is one of the reasons why patients must be completely still when having an MRI scan. The increase in the beta power of the EEG recordings indicated that the mentalist's left motor cortex was strongly activated at the time the pen fell into the glass – a split second after the "thought" was initiated by the theta and gamma waves.

At no time did my patients, myself or my siblings touch the objects that became airborne, nor did we jostle or bump into the furniture that the objects were resting on.

Based on the recorded brainwave patterns and imaging, it seems that OBEs and psychokinesis may use the left frontal and temporal lobes to bring about the ESP effects. Unfortunately, these are only tiny pieces of the jigsaw. The EEG was not recorded from the mentalist's temporal lobes and an MRI scan was not performed. Similarly, we only have half the picture for the female student performing the OBE experiment. She had an MRI scan but no EEG. These tantalising clues await elucidation with further research and investigation.

In summary, we have found that the feats of telepathy, remote viewing, OBEs and telekinesis are all associated with enhanced brainwaves in the theta and gamma frequencies for telepathy and remote viewing, and in the theta, beta and gamma frequencies with telekinesis. In other words, the brain appears to be actively engaged with these experiences.

Part IV: Distortions of Time

Now we are going to turn to something even more mysterious. Supposedly, time proceeds invariably from the past through the present to the future. This forward movement is known as the Arrow of Time. This concept of time allows no possibility to know of future events or view the past except through memories. However, a large proportion of the experiences reported by my patients were related to their ability to see or know the future – known as farseeing or clairvoyance – and to see backwards into the past, to see or sense ghosts.

18

Precognition: Seeing the Future

Precognition is derived from the Latin words *prae*, meaning before, and *cognoscera*, meaning knowledge. It is defined in the *Collins Dictionary* as "the alleged ability to foresee future events."

We will commence with the reported experiences of my patients and myself. Many were warnings of danger, either for the receiver or someone they knew. However, there were also farseeing experiences that were benign or even happy events.

Knowledge about the future while the person is wide awake

I will tell you of two of my happy experiences.

Mt Garnet is a small country town west of Cairns, which hosts the Mt Garnet Picnic races once a year. I love going to picnic race meetings. Everyone turns up: ladies in pastel high heels, matching dresses and big feathery hats; lots of little kids who run about; men dressed in singlets, stubbies and thongs. Picnic races are exciting, good natured and loads of fun.

The day my husband and I went was sunny, hot, dry and dusty. We took an esky with our lunch and set up in the pavilion. While we munched our sandwiches, I studied the race program. Suddenly – and completely out of the blue – I told my husband, "Number 5 is going to win the first race." He gave me a funny look. "Oh yeah?" he said.

Once we'd finished eating, we made our way to the paddock to look at the horses. Number 5 looked big and strong – so I put on my bet, and he won! This made me even more excited and I wondered which horse was going to win the second race. We were down among the crowd by this time and suddenly I saw in my mind a large red number two. It was like a kid's alphabet letter and had red flames shooting out of it. This seemed really weird, and I didn't trust my vision. Again, we went to the mounting yard and waited. The horses were being led around a corner from the stables and I watched three, five and one go by. The next horse was a beautiful chestnut and as she rounded the corner the sun caught her coat. It momentarily looked as though bright reddish-golden rays of light were streaming from her body. You guessed it – she was Number 2. I asked my husband if he wanted me to put a bet on for him and he declined, so he missed out on another opportunity when Number 2 won.

So it went for the rest of the afternoon, until Race 6. I found if I waited patiently – until ten minutes before the winner passed the post – the number of the horse would come into my mind. I am a very cautious gambler and only bet my original stake each time, putting my winnings away. Just as well! In Race 6, something very odd happened which I have thought long and hard about ever since.

I had received Number 4 as the winner of the sixth race and was on my way to place my bet, when Number 4 was replaced in my mind with Number 1. This was very confusing – so I stopped, looked at the horses and waited some more. Before I got to the bookie, Number 1 was replaced with Number 4 again. I was running out of time as the horses were being led onto the track, so decided I would go with Number 4. I placed my bet and made my way back to the rail. There was pandemonium on the other side of the track, as Number 4 refused to enter the starting gate three times. He got off to a bad start but caught up gamely to come second to Number 1. Number 4's antics had delayed the race by a good five minutes. I think this explains why I didn't know which horse would win – the timing was out!

Later that evening, we worked out the odds of picking the winner from the first five races. There were somewhere between five and ten horses in each race, so the probability of this occurring by chance was one in 45,000 – well past the scientific significance of one in twenty and well past the odds for a pub test.

I have dined out on this story many times since, with very mixed reactions from my audience. But there are some important inferences to be drawn from this experience which I will discuss later.

My second story isn't so dramatic but illustrates some important principles.

I was sitting at the breakfast table around 7.30am, quietly eating breakfast and letting my mind drift, when Christine popped into my thoughts. She had worked as a receptionist in my surgery about fifteen years prior. I had bumped into her two years earlier at the shopping centre and truly hadn't given her a thought since. I wondered briefly what she was doing and if she was still living in Cairns, then my thoughts drifted elsewhere.

About two hours later, I was enjoying a coffee and reading the newspaper at the airport, waiting for a flight to Brisbane. Suddenly, someone said to me, "Hi Di! How are you? Where are you going?" And there was Christine, on her way to Melbourne to see her son.

Many of my patients had similar experiences.
- Helena was enjoying her water aerobics class when she thought of a woman she hadn't seen in over ten years. Later that day, she went to the movies and there was the same woman, sitting in the theatre just a few seats away.
- Grace's aunt was sitting on her verandah having morning tea during the Second World War. Her husband was serving in the Navy. Suddenly, her husband walked around the corner of the verandah dressed in his uniform and just stood there looking at her, before disappearing. The next day he turned up as a surprise, dressed in his uniform but this time in the flesh and on leave!
- Suzanne was about nine or ten when the family living across the road brought home their new horse and tied it to one of the verandah posts. She thought this was a ludicrous place to tether a horse and thought to herself, "Wouldn't

it be funny if the horse bolted and pulled the post off the verandah?" Sometime later that day, that was exactly what happened.

All these events came true within twenty-four hours. But some patients were able to see much further ahead in time.
- Penny's grandmother was known in the family to "have the sight". On Penny's thirteenth birthday, her grandmother gave her a Bible with a white cover to carry at her wedding and said, "But of course, you won't be wearing a white dress!" Penny wore a peach-coloured dress on her wedding day, not because of any desire to fulfil her grandmother's prediction but because she had birthed a child out of wedlock to the man she was marrying. She didn't want a white wedding dress, as to her white symbolised purity and virginity, and would be wrong under the circumstances.

What principles can be inferred from these stories?
- These all happened while individuals were awake and fully conscious.
- The mind is "idling in neutral" prior to the awareness of the precognitive thought.
- The precognition comes to mind as a thought – not as a voice.
- There was a variable time delay between the event and the precognitive thought from a few minutes to a day.

Stories of foreseeing danger

Again, I have a story of my own to tell!

My husband was driving us in my car into town because I had a broken leg. It was the day after the first rain of the wet season and the roads then are notoriously greasy and slippery. He was used to driving faster around corners in his sports car than I could in my SUV. On the way in, I felt the wheels slip a fraction on one of the roundabouts and I warned him to slow down.

On the way home to the northern beaches, he was driving very carefully and less than the speed limit. About 100 metres before the first roundabout, an imperative thought came into my head: *"STOP THE CAR AND PUT IT INTO FOUR-WHEEL DRIVE."* I yelled this very loudly and urgently and he stopped and put it into four-wheel drive. I said, *"Drive VERY slowly."* We were doing no more than 20 kilometres per hour when he lost control of the car. It did a complete 360 degree turn and we ended up stopped dead in the inner lane of the roundabout – thankfully with no other traffic in sight. We could have been seriously injured! I lived in Cairns for more than thirty-five years and this was the one and only time I had such an experience.

In the days before seat belts were compulsory, my patient Jennifer had a presentiment that the car she was in with her friend was going to crash. She quickly buckled up as her friend started the car. A short time later, they came to the top of a steep hill with an intersection at the bottom out of sight. Another driver ran the red light at the intersection and smashed into their car, which was a write-off. Luckily, Jennifer escaped with minor injuries.

Errin identifies herself as a clairvoyant and volunteered that her father and sister also have the far sight. She "knows" things that are going to happen. One day she said to her sister, *"Don't go roller-skating today."*

"Why not?" the sister asked.

"Because I feel that you are going to be hurt," Errin said.

Her sister went roller-skating, fell over and broke her arm.

Errin's son Mack seems to have inherited the family's psychic abilities. He gets a vibe when something bad is about to happen, and has recently started to notice and take appropriate action. One day, he left his class to have a smoke outside the schoolyard fence. He got a bad vibe, started shaking and thought, "Get out of here right now!" Later, he discovered that the school policeman had caught fifteen students smoking, shortly after he left. His sense of danger has now saved him from trouble on several occasions.

Margaret was home in Port Macquarie, when her friend called her from a cruise ship on the Nile to see if she was okay. Her friend said she had a bad feeling about her and just wanted to check. An hour later the phone rang again. This time it was her brother, calling to tell her that their mother had died unexpectedly about an hour earlier.

John was a cane farmer and "saw" an accident unfolding. The cane harvester – an enormous machine almost two storeys tall – needed a tyre change. John "saw" the harvester slipping off the jack and toppling sideways, injuring a worker. He stopped the farm labourers and got them to chock up under the axle with timbers as a back-up for the jack. Sure enough, the jack failed, and the harvester slipped off. But no one was injured because of the precautions he had taken.

Joanna used to know if her children were in trouble. One day, when her son Luke was still a baby, she was downstairs hanging out the washing. She had an overwhelming feeling of panic, ran up the stairs and found him lying face down in his cot – not breathing. She picked him up, he took a large breath and all was well.

All the above experiences of precognition took place in clear consciousness – while the people were wide awake. In my own stories, the knowledge about the future came as a "knowing" – an absolute certainty that a certain horse was going to win, and that the car had to be in four-wheel drive.

A few people ignored the warnings they received – Errin's sister went roller-skating and broke her arm – but most people took some action that changed the course of events that they had dimly perceived. Margaret was warned, but no amount of warning would have prevented her elderly mother's death.

Mack, John and Joanna had foreknowledge of the future and all three acted, as I did on the roundabout. It seems that future events occur despite a warning, but they can be modified enough to prevent dire outcomes.

What do we know of the science of precognition?

Precognition or clairvoyance has become a field that parapsychologists have investigated with great enthusiasm over the last twenty years or so. And slowly, slowly, proof of its existence is mounting.

In 1997, parapsychologist Dean Radin explored presentiment – the unconscious emotional response to events about to occur in the very near future, or "sensing the future" – in a landmark study at the University of Nevada's Consciousness Research Laboratory.

He measured the galvanic skin response (GSR). This is a measure of the electrical conductivity of the skin, which changes with sweating of the palms and soles of the feet, as part of the body's reaction to fear or anxiety. It forms the basis of lie detector tests.

Radin's subjects had sensors attached to the first and second fingers of their left hand, to measure their GSR. They pressed a button to start recording the GSR, and to activate a computer to randomly select a photo to display on its screen five seconds later. They looked at the photo for three seconds only, but the GSR was recorded for another five seconds – making each recording a total of thirteen seconds.

The computer selected photos that were either calm – pleasant scenes and happy people – or disturbing, like graphic sex or mutilated bodies.

Usually, the GSR increases with an emotional surprise or shock **after** the person is exposed to such a thing. What Dean Radin found was that the GSR started to rise in the three seconds **before** the photo was displayed on the computer screen but **only** if it was emotionally disturbing; this did not occur if the photo was a calm one. This experiment was repeated many times with many trials per person, and the photos were chosen at random by the computer from a pool of 120 which contained forty-one disturbing images and seventy-nine calm photos.

In a later experiment, the pool was increased to 150 photos, including fifty disturbing ones, and the protocol was changed. The computer selected the photo in the instant before it appeared on the screen, rather than 5 seconds earlier. The results were the same; the GSR rose in the three seconds before the disturbing photos were displayed.[1]

This experiment has been repeated many times by many different researchers. A meta-analysis of all the data of twenty-six experiments from seven different laboratories found that the overall effect was small but stood up to statistical analysis, and therefore appears to be a true phenomenon from a scientific point of view.[2]

fMRI scans were used to see what was happening in the brain during these experiments. Direct viewing of brain activity showed an increase in the oxygen levels of the visual cortex **before** seeing a disturbing picture as opposed to a calm picture. Men reacted more strongly to erotic photos and women to both erotic and gruesome photos.[3]

Both the MRI scans and GSR measured physical changes occurring with presentiment.

Precognition differs from presentiment, in that the future is not felt but is "known."

Harold Puthoff and Russell Targ from the Stanford

Research Institute modified the protocol they used for the remote viewing experiments described in Chapter 14. The remote viewer in the laboratory completed the description and drawings **before** the site was selected. Four different locations were used, and each selected site was accurately matched by three judges to the sketches and written descriptions. This experiment clearly provides "proof" of precognition, or seeing the future.[4]

In 2011, Daryl Bem – professor of psychology at Cornell University, amateur magician and parapsychologist – took these experiments one step further by requiring the participants to think and make a deliberate choice, prior to receiving the information about the future event.

The participants selected one of two curtains displayed side by side on a computer screen. The computer then randomly selected one of the two curtains to display an erotic image – which functioned as the reward. If the participant's guess was correct, the erotic photo would be displayed, and the test would be scored as a "hit". If it was incorrect, a blank screen would show, and the test would be scored as a "miss". Hits and misses should have evened out to 50/50, but participants guessed correctly 53% of the time. This doesn't sound like very much, but it was statistically significant – it was not a chance result. A second experiment designed along similar lines scored the guess as a hit if the participant **avoided** the display of a gruesome or unpleasant photograph. This was equally successful in showing that precognition worked both with attaining a reward and avoiding a punishment.

Bem then used memory tests of word lists to see if subjects could use precognition to remember words. Unsurprisingly, people remember more of the words in a memory test which they have learned previously than words that are new. He reversed this so that words were learned in the future **after** the memory test.

Ben's subjects were shown a list of forty-eight words and were then asked to write down as many as they could remember immediately. The average correct recall was about eighteen. The computer then randomly chose twenty-four words from the original list and the subjects were given this second list and asked to sort them into groups, click on the words and finally to type them into the computer. These twenty-four words were studied in this way six times. Finally, the results of the original memory test were analysed to see if more of the words, that had been learned **after** the test, were included than the unlearned words. The results showed that this was so, therefore the subjects had used precognition in the original memory test.[5]

The publication of these experiments caused a storm of controversy in the psychology world, with some researchers agreeing with the results and others refuting them. Any new discovery is ideally investigated and tested by other researchers to see if the results can be replicated. Bem made his data, experimental design and software programs available to any other groups who wished to test the validity of his results. Many laboratories took up the challenge and a total of 12,000 participants were involved. The standout was the precognitive detection of the erotic photos. The word experiments were less successful overall. Those which used Bem's words and protocols exactly, achieved statistically significant results but those that didn't gave negative or mixed results. This meta-analysis upheld the fact of precognition.[6]

Overall, these experiments show that it is possible to measure the effects of precognition in humans, and this therefore supports the anecdotal evidence of my patient's stories. However, it is much easier to experiment with animals.

We are shifting our focus to Spain, where biologist Fernando Alvarez studied precognition in Bengalese finches.

He was aware of some physiological evidence from studies of earthworms and dogs that it was possible that animals can also "see into the future". He chose the finches because they have a very distinctive alarm behaviour in the presence of predators. They flap their wings, flick their tail to the side and call.

Alvarez placed the finches one at a time into a cage with transparent glass at two ends. One end contained a video screen showing a field of grass and the other had a view of the outside world and vegetation. The birds were in the cage for twenty minutes and were filmed with a video camera the whole time. After fifteen minutes, the video screen would show footage of a whip snake slithering through the grass in the field for fifteen seconds, at random intervals. The number of alarm behaviours were analysed in the nine second period before the snake's appearance, and compared to multiple nine second periods when the snake did not appear. These are obviously anxious little birds, as they averaged about five alarm signals in any nine second period, but an average of nine alarm signals in the nine seconds prior to seeing the snake. The statistically significant increase in the number of alarm signals just before the snake appeared showed that the finches were exhibiting presentience, like humans.[7]

What of premonitions?

Experiences of accurately seeing the future are called precognitions. Precognitions mean that someone has had prior knowledge of an incident that subsequently occurs. A premonition is something a little different.

A premonition is defined by the *Oxford Dictionary* as "a strong feeling that something is about to happen, especially something unpleasant." Unlike a precognition, a premonition can be very powerful and people can and do take avoidant action, but nothing occurs to prove or disprove that the premonition was true.

I had one of these experiences many years ago, when I was newly graduated in medicine and doing my first-year training as a doctor at Townsville Hospital.

A friend and I were sailing my boat from Forrest Beach to the Palm Island group. It was a glorious day with a gentle breeze, blue sky, the peace and quiet of the ocean – perfect sailing weather. This was shattered by my friend switching on his radio to listen to the Saturday afternoon races. By the time we arrived on the island and tied up in a mangrove-lined creek, I was feeling very annoyed and upset.

There was an inflatable canoe on board – our token effort towards a life-boat – so I hopped in with an idea of going for a row and being on my own for a while. I had only gone about two hundred metres when I started to feel very uneasy. What was wrong? Nothing that I could see; the water was brown and murky but still, the sun was shining, birds were chirping. I decided not to be silly and kept rowing up the creek. As I went on, a feeling of absolute dread came over me – so bad that I could barely put the paddle in the water. I turned back to the boat and gave up my idea of blessed solitude. I said nothing of my fear to my friend.

The following day we went snorkelling in the bay into which the creek emptied. The visibility was poor, and my friend and I kept very close together. We both decided to give up after a few minutes, again for no special reason.

Many months later, my friend and I met up in Brisbane. He confessed that when we had been snorkelling, he had felt very afraid and couldn't wait to get out of the water.

It was only a year or so later that the first crocodile attacks on people became general knowledge via the media. Nothing had happened to us, but I have always wondered if we were on a crocodile's menu that weekend. We were certainly in the right habitat.

The patients who had premonitions acted on their feelings, as I had, and changed their plans.

Karen was supposed to fly to Sydney, but she became so filled with dread that she cancelled her flights.

In their twenties, Wanda and her girlfriend went hitchhiking around NZ. They were given a lift by a man who seemed very nice. He was worried about them hitchhiking so late in the afternoon and offered to let them stay at his house. He dropped them at his home and said that he had to go and pick up his young daughter. The house was quite isolated and as Wanda walked up a rather long path to the front door, the hairs on the back of her neck stood on end. She had never experienced this sensation before, and turned to her friend and said, "I have a really bad feeling about this." Her friend said, "Me too – we have to get out of here." They ran back down the driveway to the road and eventually came to a farmhouse some distance away. They were allowed to camp for the night out of sight of the passing traffic.

Would anything terrible have happened to these people if they hadn't reacted as they did? No one knows, but each story is a clear premonition of danger. Are we seeing the future? Do people who have premonitions, but do not act, suffer dreadful fates? Again, no one knows. Is it possible that a premonition has survival value and is a part of our genetic heritage?

In the next chapter we will explore underlying science and the stories of people who "see the future" in their dreams.

# 19

Dreaming the Future

Many of the experiences my patients had, were of "seeing the future" while they were asleep and dreaming. Again, people dreamt of dangerous events happening to themselves or others. The other common theme was dreaming of places they were yet to visit.

The information contained in prophetic dreams seems to be a lot more detailed and informative than that obtained while we are awake.

Penny often "knows" of events that will occur within a ten-day period but is in the dark as to the identity of the persons involved. One day, she was discussing one of these dreams with her neighbours and said to them, "If you are in a car accident and you are upside down – **do not** release your seat belt." A few days later, their car rolled down an embankment and her neighbour found herself hanging upside down. Unfortunately, she forgot Penny's good advice, released her seat belt and injured her spine.

Barb has had many psychic experiences and related this dream. At one time, she was living in Darwin with a very abusive husband. She planned to run away from the marriage and had purchased airline tickets for herself and the children to fly to Adelaide on Christmas Eve, so they could live with her parents. One night, she dreamt that she saw her grandmother – who had died five years before – standing at the end of the bed. Barb felt scared and asked, "What are you doing here?" Her grandmother replied, "You must leave Darwin **now**; don't worry you will be okay. I will show you a photo of your next husband." She did so and then floated out the door. This dream was so powerful that Barb packed and flew to Adelaide within a day or so. On Christmas Eve, when she had planned to leave, Cyclone Tracey struck Darwin and the town was destroyed.

Elizabeth had a prophetic dream about her friends who had moved from Cairns to Speewah, on the Atherton Tableland. Her friend Janet was a schoolteacher and drove down the Kuranda Range Road daily to teach in Cairns. In the dream, Elizabeth could "see" Janet driving over the hills and valleys until she vanished from the road and then "saw" her lying very still, so that she thought she was dead. She could "see" Janet's husband, Christian, sitting at the kitchen table reading the newspaper. She was so upset by this dream that she discussed it with her friends. Janet's response was that "death may be a rebirth." Six weeks later, Christian, was reading the newspaper at the table and having a cup of tea after Janet had left for work. The phone rang. It was the school asking why Janet hadn't come to work. Christian thought, "Oh, my God – this is Elizabeth's dream!" He rushed out to look for her, but his wife and her car had disappeared. A few hours later Janet woke in her car, which had plunged over the edge of the range without any witnesses to the accident. She dragged herself out of the car and up to the road, where she was found and taken to hospital. As part of her medical assessment, she was found to be pregnant. So, not only had all the details of the dream come true, but there was a birth some months later.

Of all the stories I was told, this next one is truly remarkable.

Diana was seventeen when she had three horrific nightmares on consecutive nights, starting on a Wednesday. Each time, she ran screaming and sobbing into her parents' bedroom for comfort. In the first nightmare, there was a motor vehicle accident; she saw a dead person propped up beside a fence.

In the second dream, she was fighting with a man over an old-fashioned telephone handset, to take possession of the telephone so she could ring for the ambulance. In the third dream, she couldn't breathe because her collar bone was broken. She was struggling to pull out her collar bone, so she could take a breath.

On Saturday, Diana had an argument with her boyfriend about their regular date at the beach on Sunday. She wanted to break their date because she had a strong feeling that she had to see her friend Jenny instead. On Sunday, she rang Jenny's home to be told that Jenny had been in a car accident on Saturday night and was in Westmead Hospital. One of Jenny's friends – a sixteen-year-old – had been killed.

A week later, Jenny told Diana the story of the accident. Jenny was the only occupant of the car who was conscious after the collision. She found one of her friends sitting against the fence, but a second look revealed that she was dead. She had run to a nearby house to telephone, but the owner of the house was already calling the ambulance and she had wrestled with him to get the receiver from him. When the ambulance arrived, the officers put an oxygen mask over her face. She felt she couldn't breathe, and she was fighting the ambulance officer to try and pull the mask off.

Perhaps Diana would have seen the whole sequence on the one night if the dreams hadn't been so terrifying and hadn't woken her.

All four of these dreams came true but only Barb was able to take avoidant action, even though two of the others involved had been given specific details of the accidents. Diana did not warn her friend and she experienced the accident as if she was the victim. These stories were much more vivid and detailed than from people seeing the future while awake.

Prophetic dreams have a long history, commencing with the

Biblical story of Joseph reporting his dreams to the Pharaoh about seven years of plenty followed by seven years of famine.

John William Dunne was a British soldier, aeronautical engineer and philosopher who had his first precognitive dream in 1898 as a young man, and subsequently published a book called *An Experiment with Time* in 1928. Over the intervening years he had developed a method of logging his dreams, writing down as many details as he could remember on first waking and then comparing them with events that happened in the next two days. He discovered that he frequently dreamt of events – some trivial and some important – that occurred in this timeframe. These impacted either himself, others he knew, or strangers whom he read about in the newspapers.

He extended his research to his friends and relatives, to see if others also had this ability. They did! He then devised a theory of time which became widely known but not widely accepted. However, his ideas about his dreams are very illuminating.[1]

Firstly, precognitive dreams are of the same nature as ordinary dreams – they jump from scene to scene and event to event in juxtapositions that don't follow logic.

Secondly, the precognitive dreams would make perfect sense if they followed the night **after** the event – not the night **before**!

Thirdly, because there was a reversal of the normal flow of time, memory became entangled with time. First there was the dream, followed by the memory of the dream which was written down, followed by the event, then memory of the dream was recalled and finally there was the memory of the event combined with the memory of the precognitive dream. Dunne found that his friends and relatives **did** experience the event they had dreamt of, but would not remember the previous dream or recognise the event until they read the daily records of their dreams again very carefully.

The question that arises is: why does sensing the future not reach our consciousness? Why weren't the participants in Bem's study **aware** that the next photo was going to be emotional or not, even though their subconscious knew? It may be that the future information does not reach our awareness because most of the time it is not useful! This may explain why most of the experiences related by my patients were dramatic or life threatening; the future event may need to be very powerful to send a conscious signal. It also may explain why Dunne's friends did not remember their precognitive dreams – because they were not of any importance!

The other fascinating stories were dreams of places the dreamer subsequently visited in the real world. Again, we could say these would make perfect sense if the dreams came after the visit, and not before.

Kayla was twelve when she went with her father and sisters to a gymkhana at a place called Kerrebee Farm. She had never been there before but as her family approached the front door, she excitedly and accurately described all the buildings and their functions, as well as the layout and details of each room and its furnishings.

When Elizabeth was a young woman, she travelled to Amsterdam to join her boyfriend for a holiday. When they arrived at his address, she knew that she had been there before and was able to describe the staircase, the room he was renting and plan of the building. She remembered that she'd had a dream about six months earlier, where she had been "flying" over Amsterdam and had "seen" this particular building.

Maureen and Evelyn both dreamt of homes they were to live in.

Maureen was renting and her lease had expired. She was concerned about finding somewhere else to live but had a dream of the very house she was shown the following day by the real estate agent. She took it immediately and moved in the same day.

Eva and her husband were living in Brisbane, building their dream home over a two- year period. About nine months before it was completed, Eva had a dream "seeing" herself packing up their belongings. The voice in her dream said, "You will be moving out of your new home." One week after they moved in, her husband was made redundant, and they had to sell their house before the year was out.

Dunne was a man sixty years ahead of his time because he recognised the importance of memory in the phenomena of "seeing the future".

How the brain stores memories is one of the most important areas of research in neurology. It is now known that the hippocampus is the area of the brain where short-term memories are formed. Experiments on laboratory rats and mice have yielded a wealth of information about this process.

Researchers have inserted very fine electrode arrays into the hippocampus of rats, to measure the activity of individual neurones as the rats go about their daily business –exploring, sniffing, searching for food, feeding, grooming, napping and sleeping.

Something extraordinary happens in the hippocampus. As the rat moves through its environment from one location to the next, single neurones called "place" cells become active. The pattern of activity of the neurones corresponds to the route that the rat has travelled. The brainwaves produced by the hippocampal neurones during exploration and searching for food are gamma, locked onto theta waves. When the rat has achieved its goal it rests, feeds, grooms itself or even naps. The hippocampus then switches to a different electrical output. A special part of the hippocampus produces irregular sharp waves associated with very fast ripples or oscillations –

up to 200 Hz per second. These combination waves are called sharp-wave-ripple (SPW-R) events and last about one tenth of a second.

The information carried by the SPW-R events is a condensed version of the pattern of firing of the neurones during exploration. These signals travel almost twenty times faster than the gamma/theta waves and carry information to the adjacent entorhinal cortex – the old smell brain.

Gyorgy Bukowski and his group have studied the hippocampus in rats for over twenty years. They have found that rats "replay" the SPW-R patterns over and over while they sleep or rest, and this is the process by which short-term memories are changed into long-term memories.[2] It has been shown that these waves induce simultaneous activity in prefrontal neurones; different routes through a maze cause stimulation of different sets of neurones in the prefrontal cortex.[3] That is, the patterns are carried to the part of the frontal cortex where planning and anticipation take place.

Recently, "time" cells have been discovered. These record delays in time, as the rat runs through a maze or on an exercise wheel. So, at the end of the rat's journey, the hippocampus has stamped the route with location and time markers that represent the "where and when" of the route. This allows the hippocampus and entorhinal cortex to construct maps of the rat's environment, enabling it to remember and navigate.[4]

This is a bit like a filing system. The hippocampus catalogues the event with a date, time and place. The details of the event are then stored as a long-term memory, to be retrieved later when required.

Now we come to the interesting part.

The neuronal patterns (the sharp-wave ripples) recorded during sleep, **before** mice were allowed into a part of a maze they had never entered previously, matched the theta and gamma wave patterns that were recorded in their hippocampus while they were exploring the new section. The researchers called this "preplay". In the initial experiment, preplay occurred mostly when the rats were awake and feeding at the end of the familiar path and adjacent to the barricade at the beginning of a new path. The next experiment used a different maze that the mice had never encountered previously and again, preplay occurred in the prior rest/sleeping period.[5] Rats were later tested in a U-shaped maze with three arms. They found preplay sequences unique to each of the arms and different to each other in the sleep period before the runs.[6] Some doubt has been cast on these findings subsequently, in that some criticism has been made of the statistical methods used in the experiments.[7]

However, if we accept these findings as real, then the rats were "practising" their future explorations of new environments in their sleep or rest period **before** the event occurred. The researchers interpreted the findings as showing that the rats had various pre-existing set patterns that they could adapt to explore a new situation. Something the experimental groups did **not** consider was that preplay could be explained by precognition – the rats were imagining or "seeing" the runs ahead of time and before entering the maze. Does this remind you of my two patients – one who was able to accurately describe the internal arrangement of the rooms at the farm, and the other the rooms of the residence in Amsterdam? Likewise, of Dunne and his friends and relatives who dreamt of their experiences one or two days beforehand.

This field of study is still in its early days and hopefully one or more of these researchers will be curious enough to design experiments to test the precognition explanation.

The experimental evidence from Dean Radin's and Daniel Bem's experiments, as well as that involving the finches, backs up the ability of some humans to look forward into the future and "feel" or "know" future events or yet unvisited places.

Now that we have explored the science related to seeing into the future, it is time to consider the other end of time: looking into the past and the stories of ghosts I collected in the survey.

20

Ghosts

I have never seen a ghost but from what my patients tell me, they are not always scary. I was surprised by the number of reports of ghosts; I thought ghosts would be as rare as hen's teeth but not so!

Some people see ghosts, some hear them and some feel them. The people who "see" ghosts have often had more than one experience.

My patient Juliet was determined to walk home alone on her very first day of school because she was feeling so grown up. Her mother was reluctant but finally agreed to meet her at the top of a hill. As she was crossing the road outside the school gates, she was joined by her grandmother, who led her across the road and chatted away as they climbed the hill. As they neared the top, her grandmother said, "I have to go now, dear." Then she left. Juliet joined her mother and started to tell her all about her exciting day, including that her grandmother had met her outside the gate and helped her cross the road. Her mother was astounded. Juliet's grandmother died a few years before Juliet was born. She questioned her daughter closely, who gave a good description of her grandmother – that she wore small round spectacles and a buttoned up black coat. Juliet had never seen a photograph of her grandmother, but her description was accurate.

Juliet's mother herself "visited" Juliet about two months after she died. Juliet arrived home after a hard day's work, sat down on the sofa to chill out and plan the next day's activities.

As she got up to do her evening chores, she noticed her mother sitting beside her on the couch. "Oh, goodness!" said Juliet. "I'll just put the kettle on for a cup of tea." She had completely forgotten that her mother was dead. When she had taken two steps towards the kitchen, she remembered and turned around. But her mother had disappeared.

The third ghost to visit Juliet was someone completely unknown to her. She was working as a cleaner in a hospital in England. As she walked into a small ward, she saw an elderly lady dressed in a blue dressing gown and red slippers, leaning over the bed of a young girl. Juliet hadn't seen her in this ward before and said, "Oh, hello! I think you're in the wrong room." She turned away momentarily to assess the work needing to be done and when she turned back, the old lady had vanished. She decided not to share this experience with any of the other hospital staff, concerned that they would think her mad. Later, she overheard two nurses discussing how one of them had seen the ghost of a little old lady in a blue dressing gown and red slippers, a few days after Juliet's experience.

Maureen also had a happy experience with ghosts. She was working at a resort, which had been built on land adjoining the house where her grandparents had lived. At the time, they had both been dead for many years. The properties were separated by a creek with a bridge, connecting the two. One day, Maureen was stocking the creek with carp for the benefit of the resort guests, when she looked up and saw her grandparents standing on the bridge in the sunshine, smiling and waving to her.

Sibyll and her husband were travelling to the Atherton Tablelands via the Gillies Highway, south of Cairns. They stopped at an old pub at the foot of the mountains and as they entered, Sibyll could "feel" a woman standing on the steps that led from the foyer to the second floor. She was intrigued by this and asked the publican, "Who is the woman standing on the steps?" He said, "That's our resident ghost. She was a miner's wife whose husband was killed in a mining accident. She stood on the steps, waiting and waiting for news of him during the rescue effort."

Not all experiences are so benign.

Mark is an anthropologist who does field trips in Papua New Guinea as part of his research. A man had died in the village where he was staying, and this delayed his expedition to another village about sixty kilometres away. The villagers performed a proper burial and rituals to protect them from harm; they believed an evil spirit or person was responsible for the death. Two weeks later, Mark and a party of villagers left to travel north through a remote part of the country. The dead man had lived in this more northern area when he was young.

They travelled by motorised boat past the sago palm swamps, where semi-domesticated pigs were housed, and proceeded for about twenty-five miles upriver. They then swapped to canoes, which they paddled for the next two days. They were now in very remote and wild country. When their way was blocked by logs in the river, the men jumped out of the canoes to clear the jam. As they loosened the logs, one of them rolled over in the rushing water and the name of the dead man was seen carved into its under surface. All the shouting, joking and laughing stopped immediately and the men became very quiet and fearful. The party needed to go by foot on the last part of the journey through wild remote country. During this time, they were followed by a semi-domesticated pig for two hours. The pig was miles away from its home in the sago swamp and even more terrifying was its ear – which was cut with the clan markings of the dead man. The villagers firmly believed that they were being haunted and stalked by the spirit of the dead man, or by the evil spirit responsible for his death.

All good ghost stories are centred in "haunted" houses.

Helen and her husband had a "haunted" bedroom cupboard in their otherwise new home. During the night, the door to the bedroom where she and her husband were sleeping would slide open and the room would become very cold. Her husband told her that he saw the ghost of a little girl and said, "Can't you see her?" Helen never saw anything because she was terrified and hid under the blankets. Helen's daughter, who was four or five years of age at the time, said to Helen, "The little girl with no face who lives in the cupboard isn't going to hurt you, Mummy. Don't be scared." Helen was furious with her husband, as she thought he had discussed the ghost with their young daughter. When she took him to task, he vehemently denied talking with their daughter, but he also realised that he had never seen the little girl's face because she wore a cloak that hid it. Once they got rid of the cupboard, the "ghost" disappeared from their lives.

Susan says that psychic ability runs in her family. From the time he was a small child, her nephew always knew who was phoning and her nephew's son frequently complained that there was a man present in one of the rooms of their house. He identified the man from photographs; it turned out to be his great grandfather.

Some people only **hear** "ghosts". Often, they hear footsteps in their home.

Kayla once lived in an old house with her family. The house had lots of verandahs and an old lady had died in the home in a fire. Kayla often heard footsteps on the verandah coming to the front door, between the onset of darkness and midnight. When she opened the door, there was never anyone there. When the family were gathered around the dinner table one night, Kayla discovered that all four members of the family had also heard the footsteps.

Leanne lives in an old Queenslander style house and weird

things happen at night. If she wakes during the night and gets out of bed, she can hear people moving around and a child crying. Occasionally her bed and mattress move, as if someone has sat down on the bed beside her. Leeanne's school-age son also experiences these phenomena at night, if he is awake. She is not bothered by the noises, but her son is frightened by them.

Janice had the most fearful experiences of any of my patients. She was living in an old house in Roma – a rural town in Queensland – that had been moved from its original site on the outskirts of the town, some years previously. Her newborn baby slept in a cot in her bedroom and she woke one night to see the figure of a woman peering intently into the cot. The woman was dressed in long grey clothes, with an old grey shawl over her head and shoulders. Janice shouted at her and the woman glided over to the foot of the bed, staring towards her. Janice kept yelling, "Who are you – what are you?" There was no reaction from the apparition but eventually it left the room.

When Janice and her family moved to Mount Isa for work, they rented out the house. Her real estate agent rang to say the tenants had vacated the house and so Janet travelled back to Roma with her father. When her father entered a different room to Janet, he yelled out in a panic, "Girlie, there's something in here!"

Janice and the real estate agent were inspecting the property another day, when the apparition of a young boy dressed in a red shirt appeared in the backyard. The estate agent ran down the front stairs and Janice took the backstairs but when they reached the yard, he had mysteriously vanished. Eventually, she sold the house and moved permanently to Mount Isa.

One day, sometime later, Janice was visiting a friend, who offered to loan her some books. She picked up a book called *Up Rode the Squatter* by Hector Holthouse and published by

Angus and Robertson in 1970. The book fell open and there was a story about the Gubberamunda Ghost – a malevolent old woman dressed in grey clothing with a grey shawl, who haunted the original house which she and her husband had built in 1870, close to the town border of Roma. They lived there for a long time and then had mysteriously disappeared. Various people had been terrified by the old woman's ghost over the years and the house stood empty until it was demolished. Its timber had been used to build the house Janice had bought.

The book also told of the ghost of a young boy who had drowned on the house's property when he was about twelve. A young girl had also died of post-operative bleeding in the original house, after she had dreamt of the old woman in the grey dress and shawl telling her to remove her bandages. Thus, two of Janice's ghosts were corroborated by the author of this book.

In retrospect, Janice is very thankful to have left Roma and the house behind, but she isn't free of ghosts yet.

On relocating to Mount Isa, Janice and her family moved into an old house. One afternoon, Janice arrived home with the groceries. As she entered the kitchen, she saw her mother-in-law standing beside the refrigerator. Her mother-in-law had died a year earlier. Her daughter later came to her and said, "Mum, you are going to think I am funny but when I got home from school an hour ago, I saw Grandma standing beside the refrigerator."

They had both seen her at different times on the same afternoon.

One night, Janice woke from sleep to see a figure standing in the corner of the bedroom wearing a slouch hat and long grey trousers. She recognised it as her godfather Pat, who had died over twenty years before.

She gave her husband a nudge and whispered, "What's that

over there?" Her husband replied, "It looks like old Pat, and I've been lying here watching it for the last fifteen minutes."

It seems that some people are sensitive to the phenomena of "seeing" ghosts or visualisations of people from a past time. The person's experience was corroborated by others present at the time, or in Janice's case, by a written account which she read years after the events she described.

21

Touchy Feely-Ghosts

I have never seen a ghost, but I have felt one – I think ... Here is my experience.

My first husband, Tony, died in his early thirties of melanoma. I eventually remarried and had two boys. When they were very young, we travelled south through country New South Wales for a skiing holiday. We broke our journey by staying overnight with Tony's mother. When I walked into her apartment, I was dismayed and upset to see that she had removed all the family photos of Tony.

Otherwise, her home looked just the same – a time warp from ten years ago. I sat in the right-hand corner of the pink, cabbage-rose spattered sofa, where I always used to sit with my first husband. My mother-in-law was pottering about in the kitchen getting dinner, and my husband and boys were sitting on the floor playing.

Suddenly, I felt Tony holding my left hand. It was very strange. I could feel his knuckles, the shape of his hand, the size of his fingers and the way our palms fitted together. It wasn't a memory, it was happening – he was HOLDING MY HAND. I thought to myself, "This can't be true. If I look at my hand and see that his hand isn't there, then the feeling will stop because my eyes will be giving me a different message."

But no! I sat and stared hard at my hand but the feeling of his hand holding mine didn't go away. My distress vanished, and I felt peaceful and happy. Slowly the sensation faded away, but it had persisted for a good two to three minutes.

Roberta's son – Jack – was born with an inherited neurodegenerative disorder of the brain. He was much loved and spent his entire short life at home with the family, before dying peacefully at about six years of age. The following evening, the family was gathered around the dinner table when Roberta heard Jack laughing in his bedroom. She jumped up from her chair and ran to his room. One of her daughters arrived a second or two later. She had also heard him laughing.

Roberta was divorced from her husband Anthony, but they remained friends and on good terms until he died unexpectedly, ten years or so after Jack. One day, she plonked herself down in a chair and found herself sitting in his lap with his arms around her waist. This was one of his favourite ways of cuddling her when they were in the early years of their marriage.

The last two events involved babies.

Beverley was breast feeding Sophie in the nursery one evening when she had a very strong feeling that "someone" was in the room with them. At first, this did not worry her. But then she felt "something", as if there was a hand touching her face. She jumped up with the baby and ran into her own room. The "presence" did not seem to follow her. She was very concerned and phoned her mother-in-law for advice, who was an indigenous person of Papua New Guinea and known to be psychic,. She identified the presence as a niece who had died and visited members of the family occasionally. Beverley was to tell the presence she was welcome to watch over the baby, but she was too strong and needed to back off a bit. The following night, Beverley spoke all this into the empty room and then had no further problems.

Karen had an almost identical experience when her daughter was a baby. Karen felt another "spirit" in the room. There was also a light. She couldn't make sense of it as both her parents were alive and well, but she thought it could have been her husband's father who had died. What she did know was that the presence was benevolent.

The touchy-feely ghosts were interacting with living people, as were the ghosts of Juliet's mother and grandmother. What interests me about the other stories is that most of these occurrences seem to be glimpses of the "ghosts" going about their own affairs, without them showing awareness of their observers. Why was Sibyll's ghost standing on the staircase of the pub decades after her husband was killed in the mine? Presumably this woman eventually left the staircase, continued with her life, looked after her family and eventually died elsewhere. Was the experience of waiting for news of her husband so disturbing that she emitted a "thought wave" that is still present on that staircase and enables her identification as a very distressed woman?

Dr Michael Persinger found that he was able to induce the feeling of a "presence" – or as he termed it, a "post-mortem hallucination" – in a person placed in a rotating magnetic field of 1000 nT. The subject experienced rushes of fear and the sight of an apparition very similar to a ghost he had seen four years previously. The fear was accompanied by changes in his qEEG, showing complex spikes at 15 Hz. What is happening here? Is the presence originating as a thought pattern in the brain, or is the brain sensitised to electromagnetic radiation delivering this sensation or image from elsewhere?[1]

We need to seriously reconsider our understanding of time. How is it that we can sense the future – even if imperfectly – and see images from the past that no longer exist in our present? In the next chapter, we will explore some of the ideas about time and why it seems to be wonky!

22

What is Wrong with Time?

There is something wrong with time!
Maybe this should be: there is something wrong with "our ideas of time" or with "our perception of time." Our lived experience of time is overwhelmingly constrained by the Arrow of Time. This states that time supposedly flows from the past through the present to the future, and that our consciousness exists only in the present and proceeds from one moment to the next into the future.

Our human lives are firmly fixed by the Arrow of Time. We are born, we live and we die – in that order. No matter how often we say *if only* … we cannot change or undo the past. It lies behind us, remembered but immutable. The future expands before us, unpredictable and unknown.

If this is truly so, then it would be impossible for anyone to foresee the future, or to see or feel images from the past.

Our ideas of time

Throughout history, different civilisations and cultures have had different ideas of the nature of time. Not all believe in the Arrow of Time. Some believe in the circularity of time, that is, time is cyclical.

Physicists like Plato, Aristotle and Newton envisaged time as absolute, that there was one overall uniform "time" for the entire Universe. One second was one second everywhere.

Einstein changed this concept when he developed his "general theory of relativity", which said that time was dynamic – not absolute – but is relative, according to the gravitational field and velocity.[1] One second does not equal one second everywhere but depends on how close the measuring clock is to a large mass and the speed at which it is travelling.

Time passes more slowly when distorted by the gravitational field of a large body, such as Earth, so that clocks run more slowly at sea level than on a mountain peak. Secondly, time goes more slowly the faster a body moves through space.

It is over one hundred years since Einstein published his theories and we now have very accurate clocks in orbiting satellites that are carried at a faster speed than those on the surface of the planet. Just as Einstein predicted, time passes more slowly on the satellites and the clocks in the GPS satellites must be adjusted forwards in time, otherwise the measurement of locations on Earth would be inaccurate by a few miles. In other words, the Earth clock is showing a time in the satellite clock's future and vice versa – the satellite clock is in the Earth clock's past.

Physicist Carlo Rovelli says "time" is perhaps the greatest mystery in physics. At the most fundamental level, there is little that resembles time as we experience it. While we are free to move in space, time dictates a direction of travel, trapping us in an eternal present as it conveys us from the past we cannot revisit to the unknown future. Einstein's general theory of relativity established time as real, but the equations he used are fully reversible in time, running backwards – just as well as forwards. Relativity gives no direction to time – time just "is"! What does have direction is the second law of thermodynamics – the flow of energy as heat – so that, for example, a scrambled egg cannot be unscrambled and put back in its shell.[2]

Stephen Hawking – cosmologist, physicist and mathematician

– wrote of time in his world-famous bestselling book, *A Brief History of Time*. He attributes the Arrow of Time as being due to three different concepts: firstly, the second law of thermodynamics, in which higher energy is converted to a lower state of energy with an accompanying increase of entropy, meaning randomness or disorder; secondly, to the psychological view of time which allows us to remember the past but not the future; and thirdly, to the cosmological view of time in which the Universe is both expanding and becoming more disordered.[3]

Buddhist philosophy believes that time is an illusion and shapeless, that events co- originate but that only the mind can experience the past and foresee the future.[4] I heard a beautiful explanation of time from a young Aboriginal woman on the radio recently. *Booroogool* is the word for "the Dreaming" in her language, incorporating time – past, present and future – philosophy and morality. The old translation of this is "the Dreamtime" but a newer translation is "the Everywhen." She views time as cyclical, not as a point on a journey or as linear, but as a bubble bouncing off different things in a space.[5]

I don't pretend to understand the physics of time, space and entropy as the physicists do, but the concept of time as cyclical is not incompatible with life as we know it, as living organisms are open systems in which energy and entropy can vary – both increasing and decreasing over time.

Our perception of time

On an individual level, time is also relative; one second does not equal one second for all people and always, because our perception of time is variable. We are all aware of how quickly time flies when we are having fun and how slowly it drags when we are bored or uninterested. The problem lies with the way our brain functions and the way it measures time. Perceived time varies depending on the person's age, physiological state and brain chemistry, all of which affect our state of mind.

A journey to a new place with lots of new sights may seem to take longer than the return journey, even if the time elapsed is the same for both. Time may seem to go faster if we are motivated and interested in a task. Emotional states, such as the feeling of awe and fear, seem to induce a sensation of time slowing. Time seems to speed up with age – older people underestimate the amount of time that has passed. Very young children live in the moment and only develop awareness of time when the prefrontal cortex and hippocampus have matured. Time passes slowly for people with schizophrenia and Parkinson's disease, and those with ADHD, which affects focus and short-term memory.[6]

Everything we see happening in our world actually happened a split second in the past because processing by the visual cortex of the brain takes around a tenth of a second. Also, visual, auditory and tactile signals from the environment are processed in different parts of the brain and at different rates. Therefore, the visual cortex must wait for the slowest information to arrive, so that sound, sight and touch can be synchronised. This means that conscious awareness lags behind the timing of an event.[7] Oliver Sachs – a famous neurologist – treated patients with encephalitis, who had been immobilised and unresponsive for many years, with dopamine – similar to amphetamine. This drug is called "speed" for a good reason. The patients' brains sped up so much that one saw the flame of a match before he saw the match being struck.[8] Here, the time distortion is a matter of seconds rather than milliseconds.

Sean Harribance, the psychic, had altered time perception. His discrimination of time was of a shorter duration than that of most people. In other words, his second was shorter than most people's second.[9] Experiments have shown that the sense of time in humans can also be disturbed by an artificial magnetic field. When people were subjected to an anticlockwise rotating magnetic field, their sense of time passing slowed by three seconds or more, accompanied by an increase in theta brainwave power of thirty percent.[10]

We have internal clocks which keep track of passing time. Our body clock is sensitive to the Circadian rhythm – a twenty-four-hour rhythm regulated by the movement of Earth around the sun and the hours of daylight. The controlling clock is located in a relay centre of the optic nerve, is sensitive to light, and controls the sleep/wake cycle and all the many other metabolic clocks in the cells and organs of our bodies. Our body clock regulates the sleep/wake cycle. Sleep is vital for the body's replenishment and repair, and for long-term memory formation and learning. We are not aware of the passage of time while we are asleep due to the lack of input from the environment around us and new memories are rarely formed during sleep.[11]

Our lived sense of time depends on our memories. We cannot have a sense of the past if we can't remember it. The hippocampus in humans is responsible for stamping our experiences with a time and place, and forwarding these for processing into long-term memories in the frontal lobe of our brain during the deep sleep phase of the sleep cycle.

An unfortunate individual had both his hippocampi removed to control severe epilepsy. He was subsequently unable to form new memories, so he ended up living in a continuous present or the distant past. He would greet his carers and wife over and over, even if he had seen them many times during that day. Damage to the hippocampal neurones is responsible for the symptoms of Alzheimer's disease, with loss of recent memory and eventually most past memories as well.[12]

Time is not a feature of dreaming sleep and most people do not remember their dreams. For those who do, the dream fades very quickly from consciousness after waking. It is likely that the experiences in dreams never make it into our long-term memory under ordinary circumstances. However, if the dream is alarming or emotional, it may be remembered because it wakes us prematurely or theta power is enhanced by the amygdala so that our body and mind respond to the dream.[13]

A memory of a past event is never the same as the original experience, except for those who experience flashbacks as a symptom of post-traumatic stress disorder (PTSD), where the individual relives the actual experience many times.

It seems that our perception of time is not an absolute; it is totally reliant on our brain and its normal functioning. It can be altered or distorted by our brain physiology, our state of mind, the sleep cycle, neurological disease and our physical environment, including the magnetic field.

However, a change in the perception of time does not explain the presentience shown by the subjects of Dean Radin's experiments. Here we know that the stimulus – either erotic or disturbing images – was **not** shown to the subject before their response, as the image presentation was triggered by a random program on a computer. They were subconsciously responding to an image presented five seconds into their future.

With Daniel Bem's learning experiment, the initial testing was done minutes or hours before learning took place.

The dreams of Diana and Elizabeth were very alarming. They were remembered in great detail and described to others, days or weeks before the future events took place.

A mechanism for "seeing" the future

However, we know from the preceding chapters on precognition, ghosts and the science of these that it **is** possible for humans to look both forward in time to a future event and backwards to see or feel something from the past. Maybe, what we are accessing is **information** rather than matter.

The way I think about "seeing" the future is to compare it to the events of the Blitz in the Second World War. Initially in London, searchlights were used to try and spot and shoot down approaching aeroplanes before they could release their bombs. The plane reflected the electromagnetic wave of light back to the eye of the beholder as a visual image.

Later, radar (another form of electromagnetic radiation) was invented. This gave the defenders the ability to see the bombers from a much greater distance, so that they could warn the Londoners and launch a counter-attack. The radio waves bounced back to the radar receiver, carrying information so that the distance, speed and angle of the planes could be calculated, giving warning of a future imminent event.

If our thoughts can be carried by ELF waves at the speed of light and circumnavigate the world, they are travelling many times faster than Earth rotates on its axis and moves through the Universe. They are therefore subject to the time dilation effect caused by a difference in velocity, which slows time and theoretically brings it to a standstill at the speed of light.[14] As discussed earlier, the events occurring on Earth are then in the future of the thought wave. Maybe an "electromagnetic thought wave" can be reflected back to the brain of the sender carrying information of some future event.

John Wheeler and Richard Feynman are both famous physicists who theorised that it was possible for electromagnetic radiation to be reflected back to the present from the future. Currently, this theory is neither proved nor disproved.[15]

If this is a possibility, why are we not more aware of "seeing" the future?

The information about memories and sleep explains why

most people do not remember their dreams. During dreaming sleep, the hippocampus produces the theta and gamma waves responsible for new memories. The precognitive dreams that J.W. Dunne's friends and relatives were having were only remembered shortly after they awakened from sleep and were not processed into long-term memory. The dreams that Diana and Elizabeth had about their friends' accidents were remembered most likely due to their strong emotional impact.

It is probable that many of us have dreams about future events that are never remembered because of their unimportance. Similarly, during the day, an incoming thought signal reflected from a future event only seems to reach our awareness or consciousness when our minds are quiet and relaxed. Again, such a signal makes an impact in a life-threatening situation, such as Jennifer with her car accident, John with the tractor incident and me with the car skidding on the roundabout.

Looking back into the past to feel, see or hear ghosts seems to be more highly developed in some people than others. We all see electromagnetic radiation coming from the past. The light from a star five hundred light years away was sent on its journey at least five hundred light years ago. For all we know, that star could have become a super nova or ceased to exist during that time, but its light is still visible. It will continue past Earth and possibly exist forever, in a space in which time becomes meaningless.

It doesn't therefore seem impossible to imagine that the "thought waves" emitted by the woman on the staircase in the pub could still exist and be perceived by Sibyl. If people's thoughts can be carried by ELF waves – can the thought be so powerful that it sets up a "standing wave" that exists in the same place for many years and that this information can be accessed by a sensitive person?

It seems that the hippocampus is essential for the formation of memories. Is it also responsible for the idea of the Arrow of Time? Without a functioning hippocampus, a person becomes lost in the present and time loses its meaning. With a functioning hippocampus, we are blinded to the idea that time may not progress from the past, through the present to the future.

All these ideas about how our minds can experience the past and the future through a distortion of time are highly speculative. Physicists and others grapple with the nature and theory of time. No theory of time can be complete or accurate, if it does not allow humans and other life forms to see into – and obtain information from – the future and the past.

23

Where does God fit in?

"Where does God fit into all this?" This was the question my friend asked, after hearing about my reception of the telepathic messages of the birth of his triplets and the death of his friend.

This is a question every individual on Earth asks themselves at some stage of their life.

It seems to be a question that is a part of the human condition and is compounded with two other questions: *What is the nature of consciousness?* and *What is the meaning of life, the Universe and everything?*

When I was a very young child, I loved sunbaking and one day was lying out on the grass with my panama hat over my face. The sunlight was entering through the gaps of the weave in the hat and while I was watching the light, I was oblivious to everything else present in the world. Suddenly a terrifying thought came into my mind – "Who was I?" There was only the sun and me – so "Who was **I**?"

This is a very simple explanation of the idea of consciousness and self-awareness. This question has a group of philosophers, physicists, neuroscientists and others engaged with the "Theory of Mind"; numerous books have been written on this topic.

Where does belief or disbelief in God fit with the hypothesis that telepathy is the communication of information from one mind to another via very low frequency electromagnetic radiation, and its reception and transmission by the brain?

Religious belief, or the lack of it, is very personal and this hypothesis is open to interpretation depending on an individual's point of view. The discovery of a new sense – such as telepathy – and its modus operandi should have little influence on a person's belief system.

The atheists and agnostics who don't believe in a God should feel vindicated that telepathy and extrasensory perception have explanations rooted in the biology of the brain and in the physical laws of our Universe.

Those who have strongly held spiritual or religious beliefs may feel that the acknowledgement of an old sense and this extraordinary ability of humans is yet another of the many wonders of God and celebrate it accordingly.

My personal belief about telepathy is that it is a communication system that links and binds humanity together. If each of us can identify or know the thoughts of others who are close to us, and those others are similarly linked to yet others, then there is a widespread network of knowledge and feelings shared between humans.

I have no easily acceptable explanation for the ability of "seeing" future events or "feeling or seeing" people from the past. It appears that our perception – and ideas – of time are erroneous, because our thoughts can travel through time to gain knowledge. Maybe our brains can access electromagnetic information to give us information of future or past events.

I feel as helpless as the Vikings were in explaining thunder and lightning – which they attributed to their gods. It took over two thousand years from the time of the Ancient Greek philosophers to understand storms and the nature of electricity.

These fundamental questions are not going to be answered anytime soon – and I expect these questions will still be occupying the mind of humans for eons to come.

24

The Final Word

My aim in writing this book was to explore the phenomenon of telepathy and extrasensory perception, and to emphasise that this is a sense or ability that the majority of people living on our planet possess.

I have come to understand that many people have a belief in the truth of this sense, but there is still much fear associated with openly acknowledging this belief.

Unfortunately, the group of people most strongly naysaying this belief are the scientists and of these, the psychologists and psychiatrists are the most vehement in their opposition.

My heartfelt wish is to open the minds of people everywhere to this talent that ordinary humans possess; to remove some of the mystery and fear surrounding this old extra sense; to encourage curiosity in, exploration and investigation of extrasensory perception; and to promote its acceptance and incorporation into our daily lives.

It is a direct link to the minds of others and an adjunct to our humanity.

Appendix
Section 1: Consent, Survey Questions & Results

Method

Verbal consent

Following completion of the patient's consultation, I said to them, "I am doing some research into a topic that interests me – extrasensory perception (ESP). Would you be willing to take part in a survey about this?"

If they agreed, I then asked them three questions:

1. *"Have you or any close member of your family ever had any kind of ESP experience?"*
   This question was designed to be answered with a simple "Yes" or "No". It was deliberately open-ended and non-specific. Very few people queried what ESP meant and usually these people responded with "No".

2. *"When the telephone rings, do you ever know who is on the phone before you answer or look at the screen?"*
   Phone calls that were expected – such as a daughter checking daily on her elderly mother – or return calls from a missed call were excluded. This question was designed to see if the receiver of the call was able to telepathically detect the intention of the caller to phone.

3. *"When you phone a person that you haven't contacted for a long time, do they ever say, 'Oh, I was just thinking of you?'"* This question was suggested by a patient who had frequently been told by friends that they had just been thinking of her when she phoned, after having no contact for an interval of months or years. It seemed that this patient was broadcasting a message to the receiver, just prior to phoning.

Patients' responses were recorded on paper at the time of the interview, along with their names and the interview date. These records were kept separately from their medical files and were written, transcribed, collated and analysed by the principal researcher. They have never been sighted by any other individual. Patients' names were coded and deleted from the records prior to analysis.

The stories patients told were recorded in detail and followed up later if anything seemed unclear. The answers to the telephone questions were recorded as a simple "Yes/No", along with information about their relationship to the caller and the number of people with whom a telepathic message occurred.

The patient's medical records were accessed later, for information on age, sex and prescription of medications that may have altered their mental function at the time of the interview, such as antidepressants, opioids or anticonvulsants. This provided an approximation of how many patients had a mental or neurological illness at the time of the interview.

SURVEY STATISTICS

The survey was conducted between 8 March 2016 and 28 February 2017.

Sampling took place on twenty-five randomly selected days over this period. Five hundred adult individuals attended the practice during this period. Surveys were conducted on lightly booked days, to allow time for patient interviews. Every patient who attended on that day was eligible for entry into the survey. Persons under the age of 18 were not included.

There were 169 eligible adults, of whom 148 completed all three questions. Twenty-one patients were not included for various reasons: three were too ill, one was in a hurry, one had severe dementia, one had severe deafness, three were overseas visitors with limited English-language skills. I forgot to ask four and they did not respond to a follow-up telephone call and eight patients did not answer the telephone questions, as these were added after the first day of the survey. One hundred and fifty-six patients answered the question about ESP experience.

Results were collated and analysed with respect to gender, and the prescription of medications that may indicate the presence of a mental or neurological illness. The method of analysis used was the chi-squared (X^2) test.

(The chi-squared test is a statistical calculation which compares two samples of differing sizes, to see if the proportion of a particular attribute is more evident in one than the other. P is the probability that the result obtained will occur by chance. In biology, it is common to accept a value of P less than 0.05 (five in 100) as something that is not occurring by chance but is real. The smaller P is, the greater the likelihood that the results are real and true. If the value is greater than 0.05, the result is not significant (NS) and has occurred by chance.)

Sample size and composition by gender

The sample comprised 109 females and thirty-nine males, while the clinic population was 390 females and 109 males. The average age of the sample was sixty, plus or minus sixteen years. The sample did not differ from the clinic population in the male to female ratio:

$X^2 = 1.315$; $P = 0.252$ NS

Table 1: Sample size and prescriptions for medications for treatment of psychological or neurological illness

Scripts for Psych medications	Survey patients	Percentage	Clinic patients	Percentage
Yes	61	41.2%	133	26.7%
No	87		366	
Total	148		499	

$X^2 = 11.5309; P = 0.0006$

There were more patients with prescriptions for psychological or neurological illness (41%) included in the sample than patients with these scripts in the clinic population (27%).

Patients with psychiatric or neurological prescriptions were over-represented in the sample because they were seen more frequently, with an average of 7.86 visits over the study period, compared to 4.13 visits for patients not prescribed these medications. When the number of visits is compared between the two groups ($X^2 = 0.007$ and $P = 0.99$ NS), this discrepancy is fully explained because those with such a script were seen twice as often as those without and were therefore more likely to be included in the sample.

Of the 133 patients in the clinic population, ninety-nine had scripts for antidepressants, seven for antipsychotics, ten for anticonvulsants, three for anti-Parkinson's drugs, seven for anxiolytics and six for regular opioids. Some patients were on more than one medication – for example, an antidepressant and anticonvulsant – for treatment resistant depression or an anticonvulsant and an opioid for chronic pain.

Results

Question 1. Have you or one of your close relatives or friends ever had an ESP experience?

Seventy-two of the 156 patients included in the sample (46.2%) Had a personal ESP experience

A further twenty patients (12.8%) had a relative or close friend who'd had a personal experience. Some of these experiences were witnessed by the patient and others were told to the patient by a trusted person.

Thus, a total of ninety-two patients, or 59% of the sample, had either first-hand experience or trusted second-hand experience of an ESP event.

Table 2 ESP experience by gender

Gender	Female	Male	Total
Yes	56	16	72
No	53	23	86
Total	109	39	148

N=148, $X2 = 0.2323, p = 0.4265$ NS

There was no difference between males and females in reporting ESP experiences.

Fifty-five out of seventy-one patients with a personal experience were female and sixteen were male:

$X^2 = 1.024; P = 0.312$ NS

Seventeen out of twenty patients reporting knowledge of a second-hand experience of a friend/relative were female and three were male:

$X^2 = 1.535; P = 0.215$ NS

The gender of the person reporting either a personal or second-hand ESP experience was not relevant. Similar proportions of females and males were having these experiences.

Relationship of neurological or psychiatric medication to ESP experience

The results were analysed to determine if there was a difference between patients with a psychiatric or neurological illness, compared to those without.

Table 3: Personal ESP experience and use of psychoactive medication

	Personal ESP	No personal ESP
Medication	34	27
No medication	37	50
Total	71	77

$X^2 = 2.507$; $P = 0.113$ NS

There was no significant difference in ESP experience between patients taking prescription medications for psychiatric or neurological illness and those who were not.

Patients being treated for more severe psychiatric or neurological illness included those diagnosed with schizophrenia, bipolar disorder, epilepsy, dementia and personality disorder, along with those who were prescribed with more than one type of psychoactive medication. In total, this group comprised twenty-five patients, and was compared with the sub-sample of patients not prescribed such medication.

Table 4: Personal Experience ESP versus severe mental illness

	Personal ESP	No personal ESP	Total
Severe mental illness	15	10	25
No illness	37	50	87
Total	52	60	112

$X^2 = 2.383; P = 0.1346$ NS

Therefore, there was no association between ESP experiences and prescription of medications for neurological illness, or in those with severe mental illness.

ESP and families

Seventeen of the seventy-one respondents who had a personal ESP experience identified another family member as also having at least one of these experiences. Eight respondents identified their families as psychic, with three or more members spanning three generations having ESP experiences. In all cases, the families kept these abilities secret and did not discuss them outside the immediate family.

Results

Questions 2 and 3: Telephone questions

One hundred and forty-eight patients answered the questions about the telephone.

Twenty-four patients (16.1%) responded "No" to the telephone questions.

One hundred and fifteen persons (77.7%) responded "Yes" to at least one of the telephone questions.

Table 5: Gender and telepathy

	Telepathy: Yes	Telepathy: No	Total
Female	88	21	109
Male	27	12	39
Total	115	33	148

$X^2 = 0.49$; $P = 0.483$, NS

There was no significant difference between the numbers of males and females who responded with a yes to one at least of the telephone questions.

Receivers

Receivers were defined as patients who either knew who was on the phone before they answered, or had thought of a person who subsequently called them. Many people elaborated on the answer to these questions by saying that they would think of a friend who would then call them – most within minutes and others within a day or two.

Fifty-six of the 124 telepaths (45.2%) were receivers.

Senders

Senders were defined as those who called a person who then said, "Oh, I was just thinking about you."

Ninety-two (80%) of telepaths were senders.

That is, they broadcast a "thought message" to the recipient of the phone call.

These numbers do not add up to 124 because some of the patients could both send and receive telepathic messages.

Table 6. Telepathic sender/receiver by personal ESP experience

	ESP experience: Yes	ESP experience: No	Total
Yes	65	50	115
No	7	26	33
Total	72	76	148

n=148, X^2 =6.48, p = 0.0109 <0.05

This result indicates that those with telepathic ability are more likely to have an ESP experience. There was no significant difference between those who could either send or receive telepathic messages and those who could do both. They had almost the same number of ESP experiences.

This analysis was undertaken to see if there was a crossover between telepathic ability and personal ESP experience. This result was not significant. That is, patients with one-way telepathic communication –either sending or receiving thought messages – were no more or less likely to have had a personal ESP experience.

Table 7: Telepathy by number of persons

This analysis was undertaken to see if those who could both send and receive telepathic messages were able to communicate in this way with more than one person.

Telepathy	Send *or* Receive	Send *and* Receive	Total
>1 person	31	41	72
1 person only	36	7	43
Total	67	48	115

N = 115, X^2= 6.77, p = 0.0093, <0.01

Indeed, it was so! Seventy-two of the telepaths (62.6%) had telepathic communication with more than one individual and, those who could both send and receive telepathic messages were more likely to communicate with more than one person.

Two persons in the survey were able to receive telepathic messages from many individuals.

Simultaneous phone calls

Fourteen of the telepaths (11.2%) had simultaneous phone calls with another person. The phone would have an incoming call from the person whose number they were dialling at the same instant: they would hear the voice of the person as they commenced; or a text message would arrive as they were dialling. In the case of some patients, the same phenomena occurred with emailing or Facebook messaging.

One person shared simultaneous thoughts with her husband without the intervention of man-made communication, despite being separated and living in different houses. Interestingly, she said "No" to both the telephone questions.

Mind reading

Eighteen of the telepaths (12.1%) could read another's mind or feel another's sensations when they were in the same room or touching the person. All eighteen said "yes" to the ESP question and seventeen answered "yes" to at least one of the telephone questions. Eleven were receivers or senders and six could do both.

Section 2: The Experiences

Seventy-three of the 156 patients related 167 personal experiences.

Eighteen of the 156 patients shared an additional forty-three experiences of relatives or friends.

Collectively, grand total of 210 experiences were reported.

Table 9: Classification of experiences

Type of Experience	Personal	Others
Telepathy	61	12
Precognition	30	4
Premonition	11	1
Psychic dreams	3	0
Visualisation	6	1
Vision of dead/dying person	19	3
Sensing ghosts	13	14
Hauntings/poltergeists	8	3
OBEs	2	0
Weird things*	14	5
Total	167	43

*Weird experiences that did not fit one of the other classifications included: strange coincidences of finding information or things when needed – one patient had lost an important reference and switched on the radio to hear the topic being discussed, along with the reference; finding spare parts for repairing broken things; finding words a person couldn't spell properly; installing a fire alarm the same day her house later caught fire.

Sharing someone else's pain was common. One person had seen a UFO and another a min-min light. One patient's father was able to tell that his kids had been holding a séance in the bathroom some hours before he walked into the room.

Table 10: Results of European surveys conducted by other researchers

October 1985] *Representative National Surveys of Psychic Phenomena*

Table 9. Percentage of persons in the Human Values Survey reporting the following experiences:

	Telepathy	Clairvoyance	Contact with the Dead
Great Britain	36	14	26
Northern Ireland	24	11	12
Rep. of Ireland	19	11	16
West Germany	35	15	26
Holland	27	12	11
Belgium	18	12	16
France	34	24	23
Italy	38	38	33
Spain	20	13	16
Malta	28	18	19
Denmark	14	11	9
Sweden	23	7	14
Finland	35	15	15
Norway	18	7	9
Iceland	33	7	41
Total for Western Europe	32	20	23
U.S.A.	58	24	27

Credit: Haraldsson E. 1985 "Representative national surveys of psychic phenomena: Iceland, Great Britain, Sweden, USA and Gallup's Multinational Survey" *Journal of Society for Psychical Research* 1985 vol. p. 155

Section 3: Calculations of Word and Timing Coincidence

These calculations refer to the probability of my friend mistakenly calling me to make an appointment with the optometrist in the same five-minute interval of me thinking I needed to make an appointment for myself, and of my friend using the words "optometrist" and "appointment".

Formula: Binomial probability of success on repeated trials with two possible outcomes
$^nC_x = p^x . (1-p)^{n-x}$ where n is number of trials, x is outcome and P is the probability of x occurring

Person A calls within a five-minute period (1/12) on a day with fourteen (1/14) social hours.

(There are twelve five-minute periods in one hour and an estimated fourteen hours in the social day, for example, two people known to each other may call between the hours of 7am and 9pm if friends or relatives, as opposed to only eight business hours traditionally between 9am to 5pm.)

The probability of making a call in that particular five-minute period is 1/12 x 1/14 = 1/168.

There are two trials – one by myself and one by my friend – and the success rate is once only, so our formula becomes:
$^2C_1 = (1/168)^2 \times (1 - 1/168)^1$
= 0.000035 x 0.9961
= 0.0000348 or 35 per million

The probability of choosing the word "appointment" is 1/508 so
$^2C_1 = (1/508)^2 \times (1 - 1/508)$

$= 3.9 \times 10^{-6}$ or 4 per million

The probability of choosing optometrist is about 1/300, so:

$^2C_1 = (1/300)^2 \times (1 - 1/300)$

$= 11 \times 10^{-6}$ or 11 per million

The total chance of saying same two words in same five-minute interval is:

1.5×10^{-15} or 1×10^{-14}

Section 4

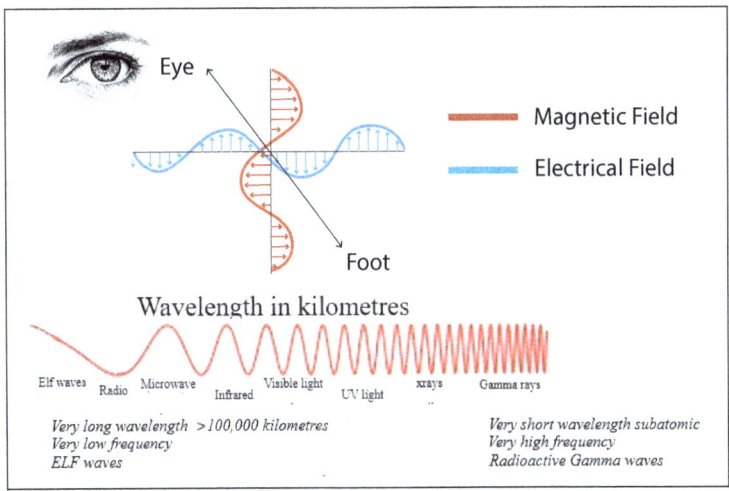

Figure 1:

The electric and magnetic fields produce electromagnetic radiation in a vertical direction. Image credit: Dr Dianne Cartwright

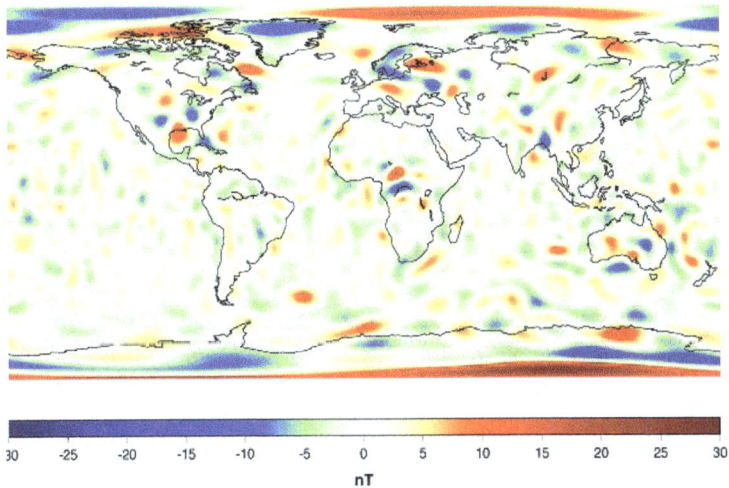

Figure 2:
Earth's Surface Magnetism. Image credit: NASA Earth Observatory

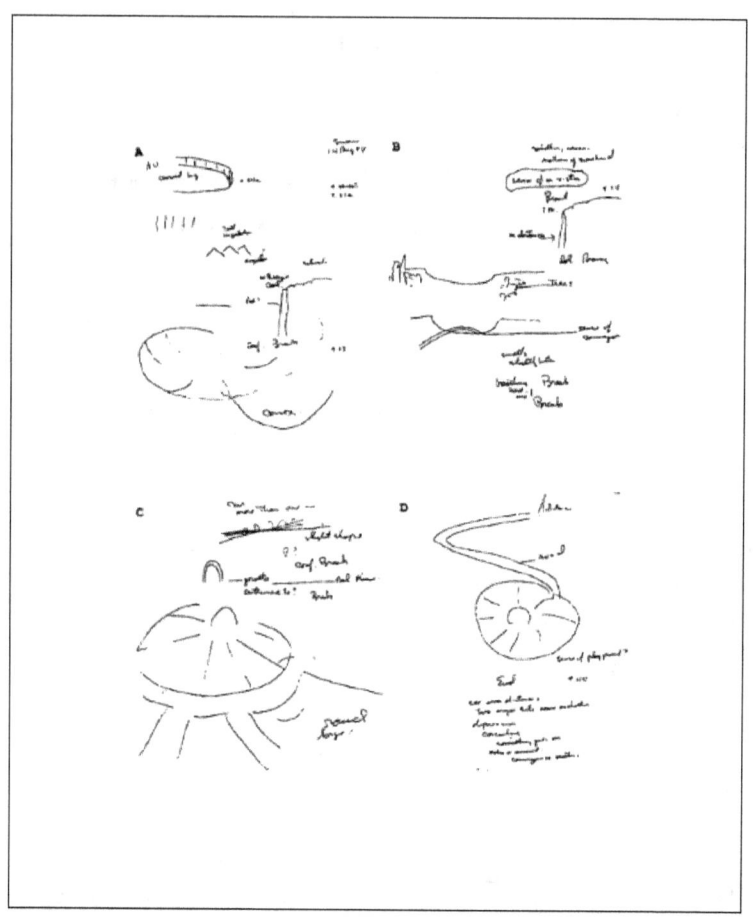

Figure 3:
Remote Viewing by Ingo Swann. Image credit: Persinger "et al 2002" https://www.semanticscholar.org/paper/Remote-Viewing-with-the-Artist-Ingo-Swann%3A-Profile%2C-Persinger-Roll/0e188cb1c0b4d692e27be7d25648d4bb07985a32

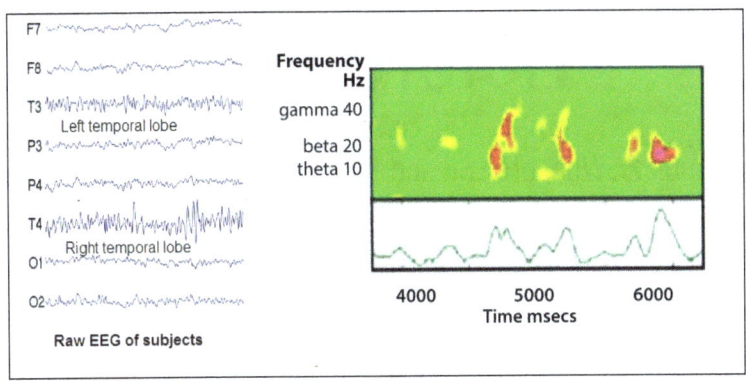

Figure 4 Spectral power showing coherence between mindreader and subject
Image Credit; Hunter et al 2010
https://www.researchgate.net/publication/277734714_Cerebral_Dynamics_and_Discrete_Energy_Changes_in_the_Personal_Physical_Environment_During_Intuitive-Like_States_and_Perceptions

Harribance's Left temporal lobe and his subjects right temporal lobe during mindreading. (Yellow= mild and red= maximum coherence)

References

Chapter 2 : What are ESP and Telepathy?

1. "Extrasensory perception", Wikipedia: https://en.wikipedia.org/wiki/Extrasensory perception
2. "Telepathy", Wikipedia: https://en.wikipedia.org/wiki/Telepathy

Chapter 3: Telepathy or coincidence?

1. Collins English Dictionary2010 https://www.collinsdictionary.com/dictionary/english/coincidence
2. Cambridge English Dictionary: https://dictionary.cambridge.org/dictionary/english/coincidence
3. Oxford Dictionary of English: *https://en.wikipedia.org/wiki/Coincidence (Stevenson Angus (2010). Oxford Dictionary of English. OUP Oxford. p. 339. ISBN 978-0-19-957112-3)*
4. "Lexical facts", *The Economist*: https://www.economist.com/johnson/2013/05/29/lexical-facts
5. "Top 1500 Nouns": https://www.talkenglish.com/vocabulary/top-1500-nouns.aspx
6. "Skilled Occupation List": https://immi.homeaffairs.gov.au/visas/working-in-australia/skill-occupation-list

Chapter 4: The survey

1. Palmer J. and Dennis M., 1974, "A Community Mail Survey of Psychic Experiences" in Parapsychology, Wiseman R. and Watt C., 2005, Taylor and Francis, NY.
2. Irwin H.J., 1993, "Belief in the Paranormal: A review of the Empirical Literature", Journal of the American Society for Psychical Research, vol. 87 (1).

3. Brankovic M., 2019, "Who Believes in ESP: Cognitive and Motivational Determinants of the Belief in Extrasensory Perception", European Journal of Psychology, vol. 15 (1). http://www.ncbi.nlm.nih.gov/pmc/articles/PMC6396695
4. Drinkwater K.D., Dagnall N., Denovan A., Williams C., 2021, "Paranormal Belief, thinking Style and Delusion Formation: A Latent Profile Analysis of Within-Individual Variations in Experience-Based Paranormal Facets", Frontiers in Psychology. https://www.frontersin.org/articles/10.3389/fpsyg.2021.670959/ful
5. Dagnall N., Drinkwater K., Parker A., and Clough P., 2016, "Paranormal Experience: Belief in the Paranormal and Anomalous Beliefs", Paranthropology, vol. 7, no. 1. https://www.researchgate.net/publication/332188015
6. Definition of rare/common medical symptoms: New Zealand Medicines and Medical Devices Safety Authority: https://www.medsafe.govt.nz:Medicinesafety
7. Haraldsson E., 1985, "Representative national surveys of psychic phenomena: Iceland, Great Britain, Sweden, USA and Gallup's Multinational Surveys", Journal of Society for Psychical Research, vol. 10. https://www.researchgate.net/publication/232443770
8. Castra M., Burrows R., Wooffitt R., 2016, "The Paranormal is (Still) Normal: The Sociological Implications of a Survey of Paranormal Experiences in Great Britain", Sociological Research Online, vol. 19 (3), p. 16. http://www.socresonline.org,uk/19/3/16.html
9. Sheldrake R., Smart P., Avraamides L., 2015, "Automated Tests for Telephone Telepathy using Mobile Phones", Explore, vol. 11 (4), pp. 310–319. http://dx.doi.org/10.1016/j.explore.2015.04.001

Chapter 8: Neurones and brainwaves

1. Naundorf B., Wolf F., Volgushev M., 2006, "Unique features of action potential initiation in cortical neurones" Nature, vol. 440, p. 1060: doi:10.1038/nature04610
2. "Lumen Biology for Majors II Action Potential". https://courses.lumenlearning.com/wm-biology2/chapter/action-potential/

Chapter 10: Coding thoughts

1. "Neural Coding", Wikipedia: https://en.wikipedia.org/wikiNeural_coding
2. "Neural Oscillation", Wikipedia: https://en.wikipedia.org/wiki/Neural_oscillation
3. Colgin, L., 2013, "Mechanisms and Functions of Theta Rhythms", Annual Review of Neuroscience, vol. 36, 2013, pp. 95–312.
4. Groppe D., et al., 2015, "Dominant frequencies of resting human brain activity as measured by the electocorticogram", Neuroimage vol. 79, pp. 223–233.
5. https://www.ncbi.nlm.nih.gov/pmc/articles/PMC4269223/
6. Zhang H., et al, 2015, "Travelling theta waves in the Human Hippocampus", Journal of Neuroscience, vol. 35 (36), pp. 12477–124876.
7. Ishii R., et al, 2014, "Frontal midline theta rhythm and gamma power changes during focused attention on mental calculation: an MEG beamformer analysis." Front Human Neuroscience, vol.; 8, p. 406. https://www.ncbi.nlm.nih.gov/pmc/articles/PMC4052629/
8. Fries P., 2005, "A mechanism for cognitive dynamics: neuronal communication through neuronal coherence", Trends in Cognitive Science, vol. 9 (10), pp. 474–480.
9. Hahn G., et al, 2014, "Communication through Resonance in Spiking Neuronal Networks", PLOS Computation Biology:

https://doi.org/10.371/journal.pcbi.103811

Figure 7 adapted from illustration in Colgin L., et al, see reference 3 above.

Chapter 11: Electrical signals bypass the normal senses

1. Hermann C.S., et al, 2013, "Transcranial alternating current stimulation: a review of the underlying mechanisms and modulation of cognitive processes." Frontier Human Neuroscience, vol. 7, p. 279. https://www.ncbi.nlm.nih.gov/pmc/articles/PMC3682121/
2. Ali M.M., Sellers K.K., and Frolich F., 2013, "Transcranial Alternating Current Stimulation modulates Large-scale Cortical Network Activity by Network Resonance", Journal of Neuroscience, vol. 38 (27), pp. 11262–11275. https://www.jneurosci.org/content/33/27/11262
3. Marquez-Ruiz J., et al, 2016, "Synthetic tactile perception induced by transcranial alternating-current stimulation can substitute for natural sensory stimulus in behaving rabbits", Nature. https://www.nature.com/articles/srep19753?draft=marketing
4. Feurra M., et al, 2011, "Frequency–Dependent Tuning of the Human Motor system Induced by Transcranial Oscillatory Potentials", Journal of Neuroscience, vol. 31 (34), pp. 12165–12170. http://www.ncbi.nlm.nih.gov/pmc/articles/PMC6623220/
5. Marshall L., et al, 2006, "Boosting slow oscillations during sleep potentiates memory", Nature, vol. 444, pp. 610–613.
6. Georgopoulos A.P., 2005, "Magnetoencephalographic signals predict movement trajectory in space", Experimental Brain Research, 25, pp. 132–135.
7. Waldert S., et al, 2008, "Hand Movement Direction Decoded from Meg and EEG", Journal of Neuroscience, vol. 28 (4), pp. 1000–1008.

8. Pais-Vieira M., et al., 2013, "A Brain-to-Brain Interface for Real-Time sharing of Sensorimotor Information", Scientific Reports – Nature, vol. 3:, p. 1319. https://www.ncbi.nlm.nih.gov/articles/PMC3584574
9. Rao R.P.N., et al, 2014, "A direct Brain-to-Brain Interface in Humans", Plos One, vol. 9 (11) e111332.
10. CIA. 2016 "Perceptual Augmentation Techniques, Part 2: Research Report", Freedom of Information Act Electronic Reading Room, CIA-RDP96-00791R000100410001-2

Chapter 12: A new sense – the magnetic field

1. Pantev C., Makeig S., Hoke M., Galambos R., Hampson S., Gallen C., 1991, "Human auditory evoked gamma-band magnetic fields", Neurobiology Proceedings of the National Academy of Sciences of the United States of America, vol. 88, pp. 8996–9000.
2. Frankel R., Liburdy R.P., 1993, in "Handbook of Biological Effects of Electromagnetic Fields Volume", vol. 2, pp. 149–183. https://digitalcommons.calpoly.edu/cgi/viewcontent.cgi?referer=&httpsredir=1&article=1298&context=phy_fac
3. Hart V., et al, and Burda H., 2013, "Dogs are sensitive to small variations of the Earth's magnetic field", Frontiers in Zoology, 10:80. http://www.frontiersinzoology.com/content/10/1/80
4. Burda H., et al, 1990, "Magnetic compass orientations in the subterranean rodent Cryptomys hottentotus (Bathyergidae)", Experienta, vol. 46, pp. 528–530.
5. Wu L. and Dickman J.D., 2011, Current Biology, vol. 21 (5), pp. 418–423. https://www.ncbi.nlm.nih.gov/pmc/articles/PMC3062271

6. Nemec P., Altmann J., Marhold S., Burda H., Oeschlager H.H., 2001, "Neuroanatomy of magnetoreception: the superior colliculus involved in magnetic orientation in a mammal", Science, vol. 294, pp. 366–368. https://pubmed.ncbi.nlm.nih.gov/11598299/
7. Gordon C., Berk M., 2003, "The effect of geomagnetic storms on suicide", South African Journal of Psychology, vol. 6, pp. 24–27.
8. Tada H. et al, 2014, "Association of geomagnetic disturbances and suicides in Japan, 1999–2010", Environmental Health and Preventative Medicine, vol. 19 (1) pp. 64–71. https://www.ncbi.nlm.nih.gov/pmc/articles/PMC3890079/
9. Rajaram M., Mitra S., 1981, "Correlation between convulsive seizure and geomagnetic activity", Neuroscience Letters, vol. 24 (2), pp. 187–191. https://www.sciencedirect.com/science/article/abs/pii/0304394081
10. Ruhensroth-Bauer G., et al, 1987, "Dependence of a sleeping parameter from the N–S or E–W sleeping direction", Zeitschrift für Naturforschung C, vol. 42 (9–10), pp. 1140–1142. https://www.ncbi.nlm/nih.gov/pubmed/2962381
11. Ravenscroft-Bauer G., et al, 1993, "Influence of the earth's magnetic field on resting and activated EEG mapping in normal subjects", International Journal of Neuroscience, 7393-4, pp. 195–201. https://www.ncbi.nlm.nih.bov/pubmed/8169054
12. Bell G.B., et al, 1994, "Frequency-specific responses in the human brain caused by electromagnetic fields", International Journal of Neuroscience, vol. 123, pp. 26–32.
13. Mulligan B., Persinger M., 2012, "Experimental simulation of the effects of sudden increases in geomagnetic activity

upon quantitative measures of human brain activity: Validation of correlation studies", Neurosciences Letters, vol. 516, pp. 54–56.
14. Kirschvink J., Kobayashi-Kirschvink A., Woodford B., 1992, "Magnetite biomineralization in the human brain", Proceedings of the National Academy of Science, vol. 89, pp. 7683–7687. https://www.pubmed.ncbi.nlm/.gov/pubmed/89:\636927dopi-Abstract
15. Dobson J., Grassi P.M., 1996, "Magnetic properties of human hippocampal tissue – evaluation of artefact and contamination sources", Brain Research Bulletin, vol. 39 (4), pp. 255–259. https://www.ncbi.nim.nih.gov/pubmed/69636927dopi-Abstract
16. Gilder S., Wack M., Kaub L., Roud S., Peterson N., Heinsen H., Hillenbrand P., Mitz S., Schmitz C., 2018, "Distribution of magnetic remanence carriers in the human brain", Nature. https://www.nature.com/articles/s41598-018-29766-z
17. Wang C., Hilbutn I., Wu D., Mizuhara Y., Couste J., Abrahams S., Bernstein S., Ayumu M., Shinsuke S., Kirschvink J., 2019, "Transduction of the Geomagnetic Field as Evidenced from Alpha-Band Activity in the Human Brain", eNeuro, vol. 6 (2), 2-19, 0483–182019. https://www.ncbi.nlm.nih.gov/pmc/articles/PMC6494972/
18. Liboff R., 2004, "Magnetic Correlates in Electromagnetic Consciousness", Electromagnetic Biology and Medicine, vol. 35 (3), pp. 228–236.
19. Hamalainen M., Hari R., Limoniemi R., Knuutila J., Lounasaamaa C., 1993, "Magnetoencephalography – theory, instrumentation, and applications to non-invasive studies of the working human brain", Reviews of Modern Physics, vol. 65 (2), pp. 414–487.
20. Caruso L., Wunderle T., Lewis C., Valadeiro J.,

Trauchessec V., Amaral J., Ni J., Jendritza P., Fermon C., Cardoso S., Freitas P.P., Fries P., Pannetier-Lecoeur M., 2017, "In Vivo Magnetic Recording of Neuronal Activity", Neuron, vol. 65, pp. 1283–1291.
21. Haraldsson E., 1985, "Representative national surveys of psychic phenomena: Iceland, Great Britain, Sweden, USA and Gallup's Multinational Surveys", Journal of Society for Psychical Research, vol. 10. https://www.researchgate.net/publication/232443770
22. Figure 1: Earth's Surface Magnetism. Image credit: NASA Earth Observatory. https://earthobservatory.nasa.gov/images/4505/earths-surface-magnetism

Chapter 13: Is the human brain an antenna?

1. CIA. 2016, "Perceptual Augmentation Techniques, Part 2: Research Report", Freedom of Information Act Electronic Reading Room CIA-RDP96-00791R000100410001-2
2. Price C., 2016, "ELF Electromagnetic Waves from Lightning: The Schumann Resonances", Atmosphere, vol. 7, p. 116: doi: 10.3390?atmos7090116
3. Kulak A., Kubisz J., Klucjasz S., Michalec A., Mlynarczyk J., Nieckarz Z., Ostrowski M., Zieba S., 2014, "Extremely Low Frequency Electromagnetic Field Measurements at the Hylaty Station and Methodology of Signal analysis", Radio Science, vol. 49, pp. 361–370. http://onlinelibrary.wiley.com/doi/10.1002/2014RS005400/abstract
4. Hamalainen M., Hari R., Limoniemi R., Knuutila J., Lounasaamaa C., 1993, "Magnetoencephalography – theory, instrumentation, and applications to non-invasive studies of the working human brain", Reviews of Modern Physics, vol. 65 (2), pp. 414–487.
5. Fernandez-Ruiz A., Muñoz S., Sancho M., Makarova J.,

Makarov V., Herreras O., 2013, "Cytoarchitecture and Dynamic Origins of Giant Positive Local Field Potentials in the Dentate Gyrus", Journal of Neuroscience, vol. 33 (39), p. 15518.
6. Wright G., Burke J., "Orthogonal frequency-division multiplexing (OFDM)", Nemertes Research. https://techtarget.com
7. Vertes R.P., Hoover W.B., Viana Di Prisco G., 2004, "Theta rhythm of the hippocampus: subcortical control and functional significance." Behavioural and Cognitive Neuroscience Reviews, vol. 3 (3), September, pp. 173–200. doi: 10.1177/1534582304273594.
8. Pape H.C., Pare D., Driesang R., 1998, "Two types of Intrinsic Oscillations in Neurons of the Lateral and Basolateral Nuclei of the Amygdala", The American Physiological Society, pp. 205–216. http://jn.physiology.org/10.220.33.3
9. "Loop antenna", Wikipedia: http://en.wikipedia.org/wiki/Loop_antenna
10. Caruso L., Wunderle T., Lewis C., Valadeiro J., Trauchessec V., Amaral J., Ni J., Jendritza P., Fermon C., Cardoso S., Freitas P.P., Fries P., Pannetier-Lecoeur M., 2017, "In Vivo Magnetic Recording of Neuronal Activity", Neuron, vol. 65, pp. 1283–1291. https://www.ncbi.nlm.nih,gov/pmc/articles/PMC6618450
11. Gilder S., Wack M., Kaub L., Roud S., Peterson N., Heinsen H., Hillenbrand P., Mitz S., Schmitz C., 2018, "Distribution of magnetic remanance carriers in the human brain", Nature. https://www.nature.com/articles/s41598-018-29766-z
12. Saroka K.S., Vares D.E., and Persinger M.A., 2016, "Similar Spectral Power Densities within the Schumann Resonance and a Large Population of Quantitative Electroencephalographic Profiles: Supportive Evidence

for Koenig and Pobachenko", PLOS One, vol. 11 (1): e0146595 https://www.ncbi.nlm.nih.gov/articles/PMC4718669/$pone.0...

Chapter 14: Interpreting telepathic information

1. Upton S., Mental Radio, David de Angelis, 2017, Digital Edition.
2. Puthoff H., 1996, "CIA-Initiated Remote viewing at Stanford Research Institute", https://www.newdualism.org/papers/H.Puthoff/CIA-Initiated%20Remote%20Viewing%20At%20Stanford%20Research%20Institute.htm
3. Persinger M., Roll W., Tiller S., Koren S., Cook C., et al, 2002, "Remote Viewing with the Artist Ingo Swan; neuropsychological profile, Electroencephalograph correlates, Magnetic Resonance Imaging (MRI) and possible Mechanisms", Perceptual Motor Skills, vol. 94, pp. 927–949.
4. Puthoff H., Targ R., 1976, "A perceptual Channel for Information Transfer over Kilometre Distances: Historical Perspective and Recent Research", Proceedings of the IEEE, vol. 64 (3): CIA-RDP96-00787R000200080046-4
5. Cage N.M., and Baars B.J., 2018, Fundamentals of Cognitive Neuroscience: A Beginner's Guide: 2nd edition, Academic Press Elsevier.
6. Aminoff E., Kestutis K., and Bar M.M., et al, 2013, "The role of the parahippocampal cortex in cognition", Trends, Cognitive Science, vol. 17 (8), pp. 379–390. https://www.ncbi,nlm.nih.gov/pmc/articles/PMC3786097/
7. Swann I., 1991, Everybody's Guide to Natural ESP: Unlocking the Extrasensory Power of your Mind, Swann-Ryder Productions LLC, sourced from Amazon Books.

8. Persinger M., 2001, "The Neuropsychiatry of Paranormal Experiences", Journal of Neuropsychiatry and Clinical Neurosciences, vol. 13 (4), pp. 515–524.
9. Venkatasubramanian G., Jayakumar P., Nagendra H., Nagaraia D., Deeptha R., Gangadhar B., et al, 2008, "Investigating paranormal phenomena: Functional brain imaging of telepathy", International Journal of Yoga, vol. 1, pp. 66–71. https://www.ncbi.nlm.nih.gov/pmx/RTICLES/pmc3144613/
10. Persinger M., Saroka K.S., 2012, "Protracted parahippocampal activity associated with Sean Harribance", International Journal of Yoga, vol. 5, 2, pp. 140–145. https://www.ncbi.nih.gov/pmc/areticles/PMC3410194/
11. Hunter M., Mulligan B., Dotta B., Saroka K., Lavallee C., Koren S., Persinger M., 2010, "Cerebral Dynamics and Discrete Energy Changes in the Personal Physical Environment during Intuitive Like States and Perceptions", Journal of Conscious Exploration and Research, vol. 1 (9), pp. 1179–119. https://www.researchgate.net/publication/277734714_Cerebral_Dynamics_and_Discrete_Energy_Changes_in_the_Personal_Physical_Environment_During_Intuitive-Like_States_and_Perceptions
12. Kobayashi M., Kikuchi D. and Okamura H., 2009. "Imaging of Ultraweak Spontaneous Photon Emission from Human Body Displaying Diurnal Rhythm", Plos One, 4(7):e6256 2009.
13. Saroka K. and Persinger M., 2011, "Detection of the Electromagnetic Equivalents of the emotional Characteristics of Words: Implications for the Electronic Listening Generation", Open Behavioural Science Journal, pp. 24–47.
14. Finn E., Shen X., Scheinost D., Rosenberg M., Huang

J., Chun M., Padademetris X., Constable R., 2015, "Functional connectome fingerprinting: identifying individuals using patterns of brain connectivity", Nature Neuroscience, vol. 18 (11), pp. 1664–1670.
15. Pape H.C., Pare D., Driesang R., 1998, "Two types of Intrinsic Oscillations in Neurons of the Lateral and Basolateral Nuclei of the Amygdala", The American Physiological Society, pp. 205–216. http://jn.physiology.org/10.220.33.3

Chapter 15: The hypothesis in a nutshell

1. DST-18108-202-78 PARAPHYSICS R&D -WARSAW PACT (U), pp. 29.
2. https://www.dia.mil/FOIA/FOIA-Electronic-Reading-Room/FileId/161532/
   Puthoff H., Targ R., 1976, A perceptual Channel for Information Transfer over Kilometre Distances: Historical Perspective and Recent Research. https://documents2.theblackvault.com/documents/cia/stargate/STARGATE%20%2311%20549/Part0001/CIA-RDP96-00787R000200080046-4.pdF
3. Persinger M. 1975, "ELF waves and ESP", New Horizons Journal, vol. 1 (5). http://www.survivalresearch.ca/NHRF/NHJ/New_Horizons_Journal_vol 5_January 1975.pdf

Chapter 16: Visitations from the dying

1. Borjigin J., et al, 2013. "Surge of neurophysiological coherence and connectivity in the dying brain", Proceedings of the National Academy of Sciences, vol. 110 (35), pp. 14432–14437. http://www.pnas.org/cgi/doi/10.1073/pnas.1308285110
2. "Near Death Experiences", Wikipedia:

https://en.wikipedia.org/wiki/Near_Death_experience#Common_Elements
3. Chawla L.S., et al, 2009, "Surges of Electroencephalogram Activity at the time of Death: A case Series", Journal of Palliative Medicine, vol. 12 (12). http://www.liebertpub.com/doi/fu;;/10.1089/jpm.2009.0159?url http://www.pnas.org/cgi/doi/10.1073/pnas.1308285110
4. Raul Vicente R., Rizzuto M., Sarica C., Yamamoto K., Sadr M., Khajuria T., Fatehi M., Moien-Afshari F., Haw C.S., Llinas R.R., Lozano A.M., Neimat J.S., Zemmar A., 2022, "Enhanced Interplay of Neuronal Coherence and Coupling in the Dying Human Brain.", Frontiers. https://www.frontiersin.org/articles/10.3389/fnagi.2022.813531/full
5. Purdon P., Sampson A., Pavone K., Brown E., 2015, "Clinical electroencephalography for Anaesthesiologists Part 1: Background and Basic Signatures", Anaesthesiology, vol. 123 (4), pp. 937–960.

Chapter 17: Brainwaves and ESP phenomena

1. Tong F., 2003, "Out of body experiences: from Penfield to present", Trends in Cognitive Science, vol. 7 (3):104–106.
2. Smith A.M., Messier C., 2014, "Voluntary out-of-body experience: an fMRI study" Frontiers in Human Neuroscience. http://www.frontierin.org/articles/10.3389/fnhum.2014200070/full
3. You tube 2014 "Stan Lee's Superheros" 10-01-2014 Kinetic Man Mind Bending
4. Brod T.M., Scott W., 2011, "Asymmetric Frontal Gamma and High Beta during Telekinesis "Demonstration". Journal of Neurotherapy, vol. 15 (4). Kindly supplied by

William Scott.

Chapter 18: Precognition: seeing the future

1. Radin D., 1997, "Unconscious Perception of Future Emotions: and Experiment in Presentiment", Journal of Scientific Exploration, vol. 11 (2), pp. 163–180.
2. Mossbridge J., Tressoldi P. and Utts J., 2012, "Predictive Physiological Anticipation Preceding Seemingly Unpredictable Stimuli: a Meta – analysis", Frontiers in Psychology, vol. 3, p. 390. https://www.ncbi.nlm.nih.gov/pms/srticles/PMC3478568/
3. Bierman D., Scholte H., 2002, "A fMRI Brain Imaging Study of Presentiment "Journal of International Society of Life Information Science 20; 380. https://scholar.google.com.au/scholar_url?url=http://uniamsterdam.nl/D.J.Bierman/PUBS/2002/fmri.presentiment.pa2002.doc&hl=en&sa=X&ei=zalMZKnbNoTmygSwu6ZY&scisig=AJ9-iYuVYUuYTOyqFfsDAkA3pYRb&oi=scholarr
4. Puthoff H., Targ R., 1976, "A perceptual Channel for Information Transfer over Kilometre Distances: Historical Perspective and Recent Research" https://documents2.theblackvault.com/documents/cia/stargate/STARGATE%20%2311%20549/Part0001/CIA-RDP96-00787R000200080046-4.pdF
5. Bem D.J., 2011, "Feeling the future: Evidence for Anomalous Retroactive Influences on Cognition and Affect", Journal of Personality and Social Psychology, vol. 100 (3), pp. 407–425.
6. Bem D., et al, 2015, "Feeling the future: A meta-analysis of 90 experiments on the anomalous anticipation of random future events", F1000Research, vol. 4: pp. 1188
7. Alvarez F., 2010, "Anticipatory Alarm Behaviour in Bengalese Finches", Journal of Scientific Exploration, vol.

24 (4), pp. 599–610.

Chapter 19: Dreaming the future

1. Dunne W., 1929, An Experiment with Time: 2nd edition, A & C Black Ltd, London. (Available on Amazon Books)
2. Buzsaki G., 2015, "Hippocampal sharp wave-ripple: A cognitive biomarker for episodic memory and planning", Hippocampus, vol. 25 (10): pp. 1073–1188. https://www.ncbi.nlm.nih.gov/pmc/articles/PMC4648295/
3. Wu X., 2014, "Hippocampal replay in a novel environment: Information content and interaction with prefrontal neuronal activities", Doctoral thesis, John Hopkins University. https://jscholarship.library.jhu/bitstream/handle/1774.2/40650/WU-Dissertation-2014.pdf?sequence=1&isAllowed=y
4. MacDonald C.J. et al, 2011, "Hippocampal 'Time Cells' Bridge the Gap in Memory for Discontiguous Events", Neuron, vol. 71 (4): pp. 737–749. https://www.sciencedirect.com/science/article/pii/S3-089662731100609X
5. Dragoi G., Tonogawa S., 2010, "Preplay of future place cell sequences by hippocampal cellular assemblies", Nature, doi:10.1038/nature09633
6. Dragoi G., and Tonogawa S., 2013, "Distinct preplay of multiple novel spatial experiences in the rat", Proceedings of the National Academy of Science of the United States of America, vol. 110 (22): pp. 9100–9105. https://www.ncbi.nlm.nih.gov/pmc/articles/PMC36707374/
7. Bendor D., and Spiers H.J., 2016, "Does the Hippocampus map out the future?", Trends in Cognitive Science, vol. 20 (3) pp. 167–169.

Chapter 21: Touchy-feely ghosts

1. Persinger M.A., Tiller S.G., and Koren S.A., 2000,

"Experimental simulation of a haunt experience and elicitation of paroxysmal electroencephalographic activity by transcerebral complex magnetic fields: induction of a synthetic 'ghost?'" Perceptual and Motor Skills, vol. 90 (2), pp. 659–74. doi: 10.2466/pms.2000.90.2.659

Chapter 22: What is wrong with time?

1. Einstein A., 2019, Relativity: The Special and General Theory – 100th Anniversary Edition, Princeton University Press, USA.
2. Brookes M., 2018, "Finding the Flow", New Scientist, 21 March 2018, no. 3174, pp. 28–32.
3. Hawking S., 1998, A Brief History of Time, Bantam Books, USA.
4. Bunnag A., 2019, "The concept of time in philosophy: A comparative study between Theravada Buddhist and Henri Bergson's concept of time from Thai philosopher's perspectives", Kaesetsart Journal of Social Sciences, vol. 40, pp. 179–185. http://dx.doi.org/10.1016/j.kjss.2017.07.007
5. Simpson N., 2020, "Nardi Simpson on Crocodile Country" on Conversations with Richard Fiedler, ABC Radio National, 13 July 2020.
6. "Time Perception", Wikipedia: http//en.wikipedia.org/wiki/Time-perception
7. Fontes R., Ribeiro J., Gupta D., Machado D., Lopes-Júnior F., Magalhães F., Hugo V., Rocha K., Marinho V., Lima G., Velasques B., Ribeiro P., Orsini M., Pessoa B., Antonio M., Leite A., Teixeira S., 2016, "Time Perception Mechanisms at Central Nervous System", Neurology International, vol. 8 (1): p. 5939. http://www.ncbi.nlm.nih.gov/pmc/articles/PMC4830363
8. Sacks O., 1986, Awakenings, Macmillan Australia.

Accessed Amazon ebooks.
9. Hunter M., Mulligan B., Dotta B., Saroka K., Lavallee C., Koren S., Persinger M., 2010. "Cerebral Dynamics and Discrete Energy Changes in the Personal Physical Environment during Intuitive Like States and Perceptions", Journal of Conscious Exploration and Research, vol. 1 (9): pp. 1179–119.
10. Cook C.M., Koren S.A. and Persinger M.A., 1999, "Subjective time estimation by humans is increased by counter-clockwise but not by clockwise circumcerebral rotations of phase-shifting magnetic pulses in the horizontal plane", Neuroscience Letters, vol. 268 (2): pp. 61–64.
11. Reddy S., Reddy V., Sharma S., 2022, "Physiology, Circadian Rhythm", StatPearls Publishing, Florida. http://www.ncbi.nlm.nih.goiv/books/NBK519517/
12. Milner B., Klein D., 2015, "Loss of recent memory after bilateral hippocampal lesions: memory and memories – looking back and looking forward", Journal of Neurology, Neurosurgery, and Psychiatry, vol. 10, p. 1136. http://jnnp.bmj.com/
13. Pare D., Collins D.R., and Pelletier J.G., 2016, "Amygdala oscillations and the consolidation of emotional memories", Trends, Cognitive Science, pp. 1364-66. https://www.cell/trends/cognitive-sciences/full text/S1364-66
14. "Time Dilation", Wikipedia: https://en.wikipedia.org/wiki/Time_dilation
15. Wheeler J., Feynmann R 1945 "Interaction with the Absorber as the Mechanism of Radiation", Review of Modern Physics, vol. 17 (2.3): p. 157.

"Telepathy our Lost Sense: Neuroscience sheds Light on ESP"

Author Information

Dr Dianne Cartwright
B.Sc. Hons (UQ) MBBS Hons (UQ) FRACGP

Dr Dianne Cartwright was born and grew up in Brisbane, Australia. She studied biology at the University of Queensland, followed by post-graduate study in the Faculty of Rural Science at the University of New England, Armidale. Later in life she returned to the University of Queensland, graduated with a medical degree, and enjoyed a long career as a GP in Cairns.

From her late teens onwards, Dr Cartwright had infrequent but fascinating and unforgettable experiences in the realm of Extrasensory Perception, including telepathy or mind to mind communication with another person. Over the years in discussions with friends, many of them confided they had similar experiences. The big question was "how many people have these experiences and abilities?"

When Dr Cartwright surveyed her patients, she was astonished to find that almost half had a personal ESP story indicating that telepathy and ESP are common and a natural part of being human. She has spent many happy and intriguing hours researching these topics during her retirement and is pleased to have found a plausible explanation to share with all those interested in and puzzled by our lost sense – Telepathy.

Find me on line at:
        diannecartwright.com.au
        hello@diannecartwright.com.au

www.ingramcontent.com/pod-product-compliance
Lightning Source LLC
Chambersburg PA
CBHW072155070526
44585CB00015B/1157